AF552474

ENCYCLOPAEDIA OF BIOTERRORISM-II

BIOLOGICAL AGENTS

By

Dr. S.K. Prasad

School of Studies of Zoology & Biotechnology

Vikram University

Ujjain

DISCOVERY PUBLISHING HOUSE PVT. LTD.

NEW DELHI-110 002

First Published-2009

ISBN 978-81-8356-381-9

Published by:

DISCOVERY PUBLISHING HOUSE PVT. LTD.

4831/24, Ansari Road, Prahlad Street,
Darya Ganj, New Delhi-110002 (India)
Phone: 23279245 • Fax: 91-11-23253475
E-mail: dphbooks@rediffmail.com
dphtemp@indiatimes.com

Printed at:

: Sachin Printers, Delhi

Preface

The present title *"Biological Agents"* aims to bring historical context to present concerns about biological weapons, biological agents, chemical weapons and the potential for bioterrorism. The lack of use of biological weapons in war advocates for using biology to create a new class of weapons initially envisioned delivery systems for pathogenic aerosols that mimicked those for chemical weapons, which were mainly bombs that generated aerosols intended to kill or disable troops in a local area. This vision was quickly replaced by the concept of creating huge clouds of germs that would drift with the wind and infect people over areas of thousands of square miles. The scientists and civil and military leaders who believed in the future of biological weapons saw their potential for fulfilling the goals of total war, for the mass killing or debilitation of enemy civilians. It is often asked why biological weapons are different from any other means of destruction. The answer is that they are the only ones devised expressly to kill defenseless humans and animal and plant life, with little real battlefield potential in modern war.

When biological weapons were developed for possible retaliation against an enemy thought to be similarly armed, they fit this model of restraint. Nevertheless, germ weapons were developed and laws and political circumstances offered no guarantees against their use. Historians may seek neat explanations, but the element of uncertainty was always there. Political and military authorities differed unpredictably when it came to calculating the consequences of using biological weapons. This book also describes the new subject of bioterrorism and traces at least the beginnings of the present era, in which domestic preparedness and homeland security are major policy issue. One of the major homeland security directives is to use technology derived from medical research to protect civilians against bioterrorism. The use of biology to defined civilians against any and all biological agents is a daunting project that imposes new security restrictions more familiar to physicists in the defense establishment than to biologists.

The study of biological weapons combines knowledge from disparate fields: biology, medicine, military history, politics, law, and ethics. This book intends to give the reader a basic literacy in this complex area. The entire subject of biological weapons is characterized by an unusual degree of misinformation and even disinformation. Almost every fact in this book has several page backstory, with greater nuance and depth than a brief overview can provide. As scholars continue their work, more information and analysis will likely turn today's accepted wisdom on its head. This progress will be a healthy sign for the field, whose subject matter has often been exploited for the frightening and sometimes entertaining effect it has on the imagination.

There can be no claim to originality except in the manner of treatment and much of the information has been obtained from the books and scientific journals available in the different libraries.

The author expresses his thanks to his friends and colleagues whose continue inspirations have initiated him to bring out this book.

The author expresses his gratitude to Mr. Wasan and staff of M/s Discovery Publishing House Pvt. Ltd. for their whole hearted co-operation in the publication of this book.

Author

CONTENTS

1

Introduction

People of most countries have suddenly had to come face-to-face with the now very real threat of the use of *biological* and *chemical* agents against them and their loved ones. The risk is smal., tiny in fact. But there is risk, and unfortunately, as recent times have so vividly illustrated, it is increasing. Everywhere people are asking what they can do to protect themselves and their loved ones from the threat of *bioterrorism*. They want to know how to cope effectively with the stress and anxiety it can cause. Here, our focus is simply on what families need to know and do to be as prepared as they can be for this threat our nation faces. Just to be clear, there is a huge difference between being pre pared and living in fear. One of the most perceptive things read by the people was of the anthrax attacks.

Recounting the experiences working in the New York offices of NBC when an anthrax-tainted letter was delivered there, Jonathan Alter wrote, "Anthrax is not contagious, but fear is." Precisely, there is no reason for paralysis in our everyday lives, and clearly no cause for panic. But there is good reason for every citizen, young and old, to know much more about what in these times might confront them. *Bioterrorism* personalizes terror like no other type of terrorism. But there are steps each of us can take to reduce our vulnerabilities and thereby restore our sense of security and safety.

What is Bioterrorism?

Bioterrorism is the intentional release of potentially deadly bacteria, viruses, or toxins into the air, food, or water supply. Ounce for ounce, *biological agents* such as *anthrax* and *smallpox* are among the most *lethal weapons* of mass destruction known. Inhalation of a millionth of a gram of anthrax may be deadly. In 1993, the Office of Technology Assessment estimated that under certain atmospheric conditions dispersion by airplane of 220 pounds of anthrax spores over Washington, D.C., could result in up to three million deaths. That's a worst-case scenario. But even on a much smaller scale, a bioterrorist attack could be designed to instill fear in our people. "As I see it, the *goal of bioterrorism* is not as much to do mortal harm as it is to disrupt our way of life and make us clearly aware of our vulnerability," says Larry Bush, M.D., infectious disease specialist at the John F. Kennedy Medical Center in Atlantis, Florida. "Though recent events have proven this successful, we will now be both intellectually and emotionally better prepared."

Bioterrorism is certainly not a new phenomenon. *Biological weapons* are as old as mankind, going back to Hannibal, who won a naval battle in 190 B.C. by firing earthen vessels full of poisonous snakes into his enemy's flagship. Recent use of *bioweapons* has only exposed and affirmed our nation's and other countries' vulnerability to these agents.

The mailing of *anthrax-laced letters* to U.S. Senate offices and media outlets last fall infected eighteen people and killed five. Tens of thousands more who were exposed to the potentially lethal biological agent were initially treated with antibiotics, and even though the attacks were confined to the East Coast, the impact could be felt across our land. But this was not the first time our nation was subjected to bioterrorism. In rural Oregon in the 1980s, members of a religious commune, in a plot to make large segments of the population too sick to vote, sprinkled *Salmonella* on the salad bars of several restaurants in the county. Over 750 people became ill. And in 1995 the Aum Shinrikyo "*Supreme Truth*" cult staged an attack with the chemical agent *Sarin* in the Tokyo subway system. The attack killed twelve people and injured five thousand. The cult also was working with bioweapons, including anthrax and botulinum toxin.

More Bioterror Events

It is no longer a question of if but when and where and how. Just as many of us never imagined that American commercial airliners would be converted into weapons of mass destruction, it is perhaps beyond the grasp of many that the weapons of choice in the first war of the twenty-first century may be *tularemia*, *smallpox*, and *anthrax*. But this should come as no surprise. The threats from biological and chemical agents are real. Terrorist groups have the resources and the motivation to use *germ warfare*. Well before the September 11 attacks, Osama bin Laden publicly stated that it was his religious duty to acquire weapons of mass destruction, including biological and chemical weapons. And when our troops went through hideouts abandoned by bin Laden's terrorist network, al Qaeda, they reportedly found evidence of efforts to develop biological and chemical weapons.

For years, we assumed that no nation or group—even terrorists—would use biological weapons because of the universal condemnation their use would surely bring from all corners of the world. Sadly, we know better now. If anything, *bioterrorism* is likely to be embraced by terrorist organizations for several reasons. As we have seen, they often are driven by fanatical religious beliefs and deep-rooted ethnic struggles. There clearly is no limit to the level of violence and death they find acceptable in support of their twisted goals. In addition, rapid advances in the technology needed to deliver biological agents have made the weaponization of germs much easier. And, with the fall of the Soviet Union, the expertise of thousands of scientists knowledgeable in *germ warfare* may be available to the highest bidder.

There are other advantages for terrorists in using a biological agent over other methods of attack. It is very difficult to trace a *bio-agent back* to its source. In general, neither Sophisticated knowledge nor significant resources are required to launch a bioterrorist attack, and the materials are relatively easy to acquire. It has been estimated that a substantial *biological arsenal* could be developed in a fifteen-foot-square room with just $10,000 of equipment.

Biological weapons pose considerable security challenges that are different from those of standard terrorist devices. They are not detected by methods used for explosives and firearms, such as metal detectors and x-ray devices. They are invisible. The first indication of an attack is the development of symptoms by those exposed. Yet, victims of a covert bioterrorist attack do not necessary develop symptoms immediately upon exposure to the bioagent. Symptoms may occur days later, long after the *bioweapon* is dispersed.

As a result, people who are exposed will most likely arrive in emergency rooms, physician offices, and urgent-care centers or clinics with symptoms that mimic the common cold or flu. In all likelihood, physicians and other health care providers will not attribute these symptoms to a bioweapon. If the bioagent is *communicable*, such as *smallpox*, many more people—including health care workers—may be infected before doctors know what they are dealing with.

"High Alart"

After the September 11 attacks, the federal government issued several notices placing law enforcement agencies and the military on "*high alert*" when credible information of possible terrorist attacks had been compiled by intelligence sources. At the same time, general alerts were issued to the public.

The alerts are a way for the federal government to let citizens know that the military and law enforcement agencies are increasing their vigilance and that citizens should, too. I know people feel frustrated because they don't know exactly what to do. It can be stressful for them and their families.

Frankly, our government is in a tough spot on this one. If we receive what we believe is credible information regarding a possible terrorist attack—even if key details such as when and where are missing—shouldn't the government let the people know? But is there really anything you can do to help? Absolutely. First and foremost, you can be the eyes and ears of our law enforcement agencies. You know your communities better than anyone else. You know when something looks out of place, whether it's a package left on the subway or someone acting in an unusual or suspicious manner in your neighbourhood.

Being more *vigilant empowers* you to be part of our war to rid the world of the evil of terrorism. Be more conscious of what's going on around you. Report any suspicious activity or behaviour to local authorities. But vigilance alone is not enough. You should take additional steps now to plan and prepare for how your family will respond if there is a bioterrorist or some other form of attack.

Selection of a Place

Terrorists tend to choose highly visible targets where large numbers of people gather. These would include large cities, international airports, subway systems, resorts, historic landmarks, and major sporting and entertainment events. It's not that people should avoid these places. In fact, it is important that we not give in to fear by allowing the terrorists to change the way we live. Instead, whenever you are in one of these situations, just be a little more aware of your surroundings.

Learn where the emergency exits and staircases are. Plan ahead how you would get out quickly in an emergency. If you are traveling, take note of any conspicuous or unusual behaviour. Don't accept packages from strangers, and never leave your luggage unattended.

What to Do in Case of Biological or Chemical Materials?

Don't panic! Yes, every situation is different, but there are general steps that will minimize risk to you and your loved ones. And these apply to both biological and chemical events.

1. If you are outside, evaluate the suspected area from a position upwind cover all exposed skin surfaces, and protect your respiratory system as much as possible perhaps using a handkerchief to cover your mouth and nose.
2. If the incident is inside, leave immediately and try to avoid the contaminated area on your way out. Keep windows and unused doors closed. Turn off the ventilation system (air conditioning or heat). If you are inside and the event is out side, stay inside. Turn off the ventilation system and seal windows and doors with plastic tape.
3. Call 100 or 911 and report the following:
   - Your name and phone number
   - Date and time of event
   - Distance from the incident or point of impact
   - Reason for the report (for example, people becoming sick, a vapour cloud, dead or sick animals or birds, unusual odors, dead or discoloured vegetation)
   - Location of the incident
   - Description of the terrain (for example, flat, hills, river)
   - Weather
   - Temperature
   - Odor (e.g., none, sweet, fruity, pepper, garlic, rotten eggs)
   - Visible emission (for example, none, smoke, haze)
   - Symptoms (for example, none, dizziness, runny nose, choking, tightness in chest, blurred vision, fever, difficulty breathing, stinging of skin, welts/blisters, headaches, nausea, and vomiting) and time they appeared
   - Explosion (for example, none, air, ground, structure, underground) and location
4. Once clear of the suspected contaminated area, remove all external clothing and leave it outside. Proceed directly (within minutes) to a shower and thoroughly wash with soap and water, scrubbing aggressively to cover every part of your body with at least ten scrubbing motions. Irrigate your eyes with water.

Kind of Plan

Every family should have a disaster plan. If yours does not, start discussing a plan tonight at the dinner table. Even without the threat of bioterrorism, this is a sound idea. The current world situation only reinforces the need for preparedness. We are not talking about *bomb shelters* here or the *pre-Y2K hysteria* to which some fell prey. The things you should do to safeguard your family in case of a bioterrorist attack are basically the same as what you would do for any natural disaster. Your plan should cover three essential elements, according to the Red Cross:

1. *Communication*: How will you communicate with family members if there is a bioterrorist attack or some other disaster?
2. *Destination*: Where will your family go if there is an attack?
3. *Supplies*: What supplies should you have on hand in case you need to "*shelter at home*" for a while?

Kind of Communication

Choose one person who lives out of state to be your family's contact in case of emergency. Why? In a disaster, it is often easier to call long distance than to make a local call. Everyone in your family should know the address and phone number. Also, choose a family meeting place outside your neighbourhood in case you can't go home. Again everyone in your family should know the address and phone number. In addition, you should have a backup out-of-state contact and a backup meeting place just for insurance. And be sure to discuss your plan routinely with family members so that it becomes second nature. That will help prevent panic if disaster does strike.

Safe Room

That obviously will depend on how close the attack occurs to your home. The most likely scenario appears to be that emergency officials would urge people to shelter at home in the event of a bioterrorist attack. So designate a "*safe room*" in your home, one with a telephone and radio. Choose an interior room without windows, if possible. Don't use the basement, however, because a chemical attack—heavier chemical vapours would tend to sink to the lowest place in a house. Gather your family in the safe room and listen to the news for further instructions. If officials order an *evacuation*, make sure everyone in your family knows in advance how to get outside from every room in the house. Where possible, devise two escape routes from every room, in case one is blocked.

Disaster Supply Kits

Disaster supply kits are Just what they imply: a collection of basic supplies that are readily available in the event there is a "*worst case scenario*" that requires you and your family to become fully self-sufficient for several days. In the case of bioterrorism, this might occur because stores are closed and other social services interrupted. The disaster kit for bioterrorism is not very different from that required for other types of emergencies.

Pack essential supplies in an easy-to-carry container such as a large, covered trash can, a duffel bag, or a camping backpack. Store this disaster supply kit where it can be easily reached, and make sure every family member knows where it is. That way, you can grab it quickly whether you have to remain inside or evacuate.

Your disaster supply kit should include the following items:

1. *Water*. Store in plastic containers such as large soft-drink bottles. Have at least a three-day supply, figuring on one gallon a day for each person. Change the water in your kit at least every six months.
2. *Canned food*. At least a three-day supply. Good items include canned meats, fruits, and vegetables; juices, milk, and soups; high-energy snacks such as peanut butter, jelly, crackers, "*power bars*," and trail mix; candy and cookies; instant coffee and tea bags; and any special foods for infants, the elderly, or those on special diets. Avoid salty items, though, as they will make you drink more water. Write the date on the food items and change foods at least every six months. Be sure to check expiration dates on labels. An easy way to remember to update your water and food supplies is to change items at the beginning and end of daylight savings time, when you also change the batteries in smoke detectors.
3. Non-electric can opener.
4. Cell phone.
5. *Change of clothing*. One extra set of clothes and footwear for each member of the family.
6. *Goggles*. One pair for each member of the family, to protect the eyes. Swimmer's goggles are fine.
7. Respirators for each family member. These are filtered fiber masks and only cost about $1 each. Look for ones with N95 certification.
8. Roll of plastic tape, such as package tape (to seal windows, if necessary).

9. Flashlight with extra batteries.
10. Portable radio with extra batteries.
11. Portable heater.
12. Thermal blankets or sleeping bags. One for each member of the family.
13. Extra car keys, credit card, and cash.
14. Extra pair of eyeglasses.
15. Special items for infants or elderly or disabled family members.
16. First aid kit, including:
    - A ten-day supply of your family's prescription medications
    - Painkillers, such as *ibuprofen*, *acetaminophen* or *aspirin*
    - *Antihistamines* if family members have allergies
    - Mild laxative
    - Antidiarrhea medication such as Pepto-Bismol Kaopectate, or Imodium AD

    If you have pets, include the following items in your supply kit:

    - Identification collar and rabies tag
    - Carrier or cage
    - Leash
    - Medications
    - Newspapers, litter; trash bags for waste
    - Two-week supply of food and water
    - Veterinary records (necessary if your pet has to go to a shelter)

Best Protection

The best bet is a simple mask with a filter that covers your mouth and nose and can block extremely tiny particles, The technical term for this kind of mask, manufactured by 3M as well as other companies, is a "respirator." Look for one with the rating of N95.

There are generally three categories of masks: ones that enclose the entire head ("*full-face*"), ones that cover the mouth and nose ("*half-face*"), and the simple disposable masks where the mask itself selves as a filter. If fitted correctly—and this is harder than it sounds—respirators can reduce exposure to anthrax bacteria and other harmful agents. It is very difficult, however to achieve a perfect fit with these simple masks. It takes about fifteen minutes to adjust the mask to fit your face the right way. So just owning one, without a proper fitting, may provide a false sense of security.

Now, about the rating. Filters are tested by the National Institute for Occupational Safety and Health (NIOSH). The rating reflects how efficiently the filter blocks tiny microscopic test particles of 0.3 microns in diameter. A filter with a 95 rating filters out at least 95 percent of the particles in a test. A filter with a 100 rating filters out 99.7 percent of the particles.

Filters are also classified into one of three categories: N, P, or R. *N* means the filter was tested with sodium chloride particles. *P* signifies that the filter is impenetrable by oils. *R* signifies that a filter is resistant to oils. So you need to understand that N95 masks only reduce exposure to particles, they don't eliminate it entirely. Since at this point we still do not know exactly how many anthrax spores, for example, it takes to infect a person, there's no guarantee that wearing a mask would offer complete protection—even if it screens out 95 percent of the particles.

The majority of filters you'll hear about are N95, P100, or N 100. All N95 through P100 masks should be fit tested. The Occupational Safety and Health Administration (OSHA) does not require fit testing for disposable N95 masks if they are worn voluntarily. You should be aware that any facial hair where the mask meets the skin greatly reduces the efficiency of the mask.

A more expensive option is the rubbery half-face mask that has filters protruding on the sides. It's called an "*elastomeric respirator*," in industry terms. *Elastomeric masks* have replaceable filters that screw on and off, and the synthetic material of the mask can be disinfected with cleaning material. The P100 elastomeric mask does the best job of protecting against biological agents.

Masks Costing

Prices for masks vary. Disposable N95 respirators are about $1 each. N100 disposables are around $5. Disposable respirators are available at most hardware or paint stores. Available at hardware or industrial-supply stores, elastomeric masks are about $20 each, and filters can be purchased for $5 and up, depending on the efficiency of the filter. All NIOSH-approved filters are marked as such on the box, the filter itself or in the instructions—look for the NIOSH approval.

The less expensive but widely marketed "*dust and nuisance*" masks are not NIOSH approved and are essentially ineffective for biological agents. But they are useful for visible dust.

The top respirator manufacturers are 3M, Moldex, and MSA.

Filtered Masks

None of the *industrial masks* mentioned above are effective against chemical agents. Chemicals are effectively filtered by using activated-carbon filters that are treated with impregnate. Different filters are typically made for specific chemicals. However, some manufacturers make carbon filters that protect against several different chemicals.

Military-type gas masks that protect against chemicals are not rated by NIOSH for use because they physically restrict breathing and could harm the general public. Highly sophisticated "supplied- air" respirators that deliver their own self-contained clean air are effective for most chemical exposures. However, no respirator is capable of preventing all airborne contaminants from entering the wearer's breathing zone. A mask is ineffective against contaminants that can penetrate exposed skin, such as sulfur mustard gas.

Within a year, NIOSH and the U.S. Department of Defense will for the first time come up with specific approval standards for protection against chemical and biological weapons.

Simple Fiber Masks

Fiber masks without filters are simple masks that cover the mouth and nose, similar to those you see surgeons wearing on television. The main purpose of these simple masks is to keep out dust and other particles that may fill the air in disasters such as the World Trade Center attack. Fiber masks without filters will not be very useful in a biological or chemical attack because the mesh of fiber masks is not small enough to keep out the very tiny biological or chemical agents. Only *gas masks* and other specialized industrial masks will work in such situations, and even they would have to be worn at the time of the attack. Putting a mask on after you were exposed would be too late.

Therefore, is recommend that you include in your disaster supply kit an N95 NIOSH mask for each of your family members. Without a personalized of however this mask offers only limited protection against biological agents than a simple fiber non-filtered mask, but not as complete as industrial grade respirators.

Gas Mask

The interest in *gas masks* when faced with the threat of biological or chemical attack is certainly understand especially when you learn that so many of the *biological agents* are so deadly when they are inhaled into the lungs.

True *industrial-grade gas* masks can protect against biological as well as chemical agents. A sophisticated, powered, air-purifying respirator with high-efficiency particulate air (HEPA) filters, when properly fitted, can reduce inhalation exposures by 98 percent.

But—and this is a very important point—this kind of mask is effective only if you are wearing it at the time of the attack.

If there is a biological or chemical attack, one thing is virtually certain: It will come without warning. And that's really the key issue here, in terms of the practical question about buying a gas mask. Unless you have enough advance warning to get the mask, put the right filter in, and make sure it's fitted properly, it won't do you any good.

And the odds of that happening are slim. In an urban setting such as New York, sheltering in your safe room and immediately turning off the ventilation system (heat or air conditioner) is likely to be more useful than a gas mask.

Decontaminate an Area

Although your natural reaction might be to immediately *decontaminate* an area, check with your local public health authority first, because you may be putting yourself at risk by further exposure to the offending agent during the cleanup. Moreover, you may lose important information and evidence that could be useful to a law enforcement investigation.

For suspected biological and chemical decontamination, the contaminated areas can be washed with a 0.5 percent sodium hypochlorite solution, allowing a contact time often to fifteen minutes. To make a 0.5 percent sodium hypochlorite solution, take one part household bleach, such as Clorox, and ten parts water. Keep this solution away from your eyes.

If biological or chemical contamination is suspected on fabric, clothing, or equipment you can decontaminate with undiluted household bleach. Leave it on for thirty minutes before discarding or using the item.

Most Terrifying Event

First, realize that you are not alone! A nationwide study showed that nine out of ten Americans showed some clinical sign of stress the week after the *September 11 terrorist attacks*. And almost half of all adults reported at least one substantial symptom of stress, such as having difficulty sleeping or uncalled-for outbursts of anger.

We can learn much from others around the world. Citizens in Israel and Northern Ireland have been living under the threat of terrorism for years. They have experienced the terror but have accommodated and adapted and been able to regain normalcy in their lives.

We know from research that while the effects of a disaster such as the terrorist attacks typically lessen with time, they can linger for years and may resurface from time to time.

So what are the signs that you are continuing to feel stress over the terrorist threat?

Reactions may be physical or emotional. Among the physical symptoms are backaches, headaches, stomachaches, diarrhea, nausea, and shortness of breath. Emotional symptoms include shock, disbelief, fear, grief, anxiety, disorientation, hyperalertness being easily startled, nightmares, crying, anger, irritability, detachment, numbness, feelings of betrayal, survivor guilt, isolation, depression inability to concentrate or carry on normal activities, a sense of loss of control, revival of past traumatic memories, apathy, and decreased ability to feel joy.

If you are experiencing any of these signs, know two things You are not alone. And there are simple steps you can take to feel better. Here's what mental health experts recommend to relieve stress and anxiety:

Talk with Others

This is a time to draw closer to those you love and trust, not push them away. And chances are they are feeling many of the same things as you. Just giving voice to what you are feeling inside can lessen stress.

Keep the Faith

For many of us, faith was a source of comfort and strength in the wake of the unspeakable horror we witnessed. As a medical doctor, I know the healing power of prayer. In these difficult times, prayer can help ease anxiety and bring us together. This is a time to draw strength from the traditions of your church, synagogue, mosque, or temple. Knowing that God is just and that he is in control offers great comfort when we feel so powerless.

Embrace your Daily Routines

It's important for all of us to go about our lives as normally as possible. If you are feeling like your life is spinning out of control, the simple, productive daily routines—whether it is taking a walk, listening to an audio book on your commute to work, or shuttling the kids to soccer practice and music lessons—offer calming reassurance that life does go on. In particular, make a point of doing the things you do well. Whether it's baking cookies, playing basketball with your daughter doing a crossword puzzle, crocheting a scarf—whatever, it doesn't matter. The idea is to feel once again that sense of being in control and excelling at something.

Take a News Break

Television news can become seductive. It is important for all of us to be informed. But with twenty four hour cable news networks, the Internet, and a huge assortment of newspapers and news-magazines it's easy to feel overwhelmed by too much information. And this can heighten your anxiety. In the week after the September 11 attacks, adults and children watched on average an hour more of television a day, and those who watched the most TV news coverage exhibited the most signs of stress. Again, this is not to say that watching the news, reading the newspaper listening to the radio, or checking out your favourite news sites on the Web is harmful in any way. It is a matter of being selective about how much news you get. If your news habits are feeding your anxiety, it's time to cut back.

Show your Colours

We saw a rebirth of overt, good old-fashioned patriotism after September 11. And that is a magnificent and healthy reaction. The greatest thing about the people is our indomitable spirit. The outpouring of generosity and caring this country witnessed after the September 11 attacks. Seeing yourself as part of this great land is more important now than ever before. If you are feeling isolated or not in control, small acts can help. Wear a flag lapel pin. Our family put up a large permanent flagpole in our yard so we could fly the American flag daily. Send a donation to a local charity. Get involved in your community's preparedness plan. Do what you can to support your country during these trying times.

Join a Group

A sense of community, of belonging to something larger than yourself goes a long way in overcoming feelings of isolation. Shared interests can help you expand your network of friends. So join a book group take a modern dance class, sign up to study a foreign language—again, whatever interests you, whatever you have always wanted to do.

One of the encouraging signs from the nationwide study on post—September 11 stress is how many Americans turned to these coping strategies to deal with the anxiety they felt. According to the study, which was done by the Rand Corporation and the University of California at Los Angeles, 98 percent of those responding to the survey said they talked with others to help themselves cope; 90 percent said they turned to religion; 60 percent said they participated in group activities; and 36 percent said they made donations to relief funds.

Also, make sure you take proper care of yourself. Here are some other ways that mental health experts say you can relieve stress and anxiety:

Exercise Regularly

During stressful times, exercise often is the first thing to go. You may not feel like you have the time or energy to exercise. But that's precisely when you need it the most. Exercise can actually boost your energy and give you a greater sense of control. Going to the gym is terrific, but there are lots of other ways to fit exercise into your hectic life. Start small. Walk with a coworker, friend, or family member at lunchtime. Take family hikes or bike rides on weekends. The key is to make exercise part of your routine. Just avoid exercising within three hours of bedtime, so it doesn't interfere with needed sleep.

Eat Well

Healthy eating is another of the first things to go during stressful times, usually for the same basic reasons as exercise: lack of time and energy. But just like exercise, eating regular meals with healthful foods is a great stress buster. Make sure you eat five to nine servings of fruits and vegetables daily, and avoid too much caffeine, sugar, and alcohol.

Get a Good Night's Rest

Tempting as it might be, turning to alcohol or relying too much on prescription or over-the-counter sleep aids can have a negative effect on your ability to get restful sleep. If you are having trouble sleeping, eliminate caffeine from your diet, talk or write about your concerns and worries before you try to sleep, and avoid late- night news or other activities that get you agitated. Take a warm bath. And, yes, drinking warm milk does help some people sleep.

One last note: Everyone copes with tragedy in different ways. There is no one right way. So be willing to experiment and find out what works for you. Recognize that it may take time before the stress and anxiety you feel subside. That is perfectly normal.

Traumatic Experiences

People are often surprised that reactions to *trauma* last longer than they expect. Resilient as we are, it may take weeks, months, or, in some cases, years to regain equilibrium. *Traumatic experiences* often remain vivid forever. Most of us will always be able to say where we were when we first learned of the terrorist attacks and the events that followed.

Some people will work things out with the help and support of family, friends, and colleagues. But these days, many are finding that they or their children have a number of stress reactions that won't go away or that interfere with work, school, home life, or leisure activities. Talking to a mental health trauma specialist can help. The key is to reach out—ask for help, support, understanding, and opportunities to talk.

Greater Risk for Certain Types of Bioterrorism

They may be, and this is an area to which we need to devote a lot more research. Certainly for *chemical weapons*, this is true. Nerve agents, including Sarin—are denser than water, so they concentrate closer to the ground, in the breathing zone of a child.

Children breathe faster than adults, which would potentially place them at greater risk for inhalational agents. They are more likely to have minor cuts and scrapes, which could make it easier for germs to enter the body. For their body weight, children have a greater skin surface than adults, which could expose them more to blister agents, such as mustard gas. In addition, many of the potential bioterror agents cause vomiting and diarrhea, which can quickly lead to dehydration and shock in children.

To further complicate matters, the vaccines and antibiotics to fight these biological agents have generally not been studied as well in children as they have been in adults. The anthrax vaccine, for instance, has never been licensed for persons under eighteen.

But we in government are working to change this. Age-related research is critical. As new antibiotics and vaccines are approved, we must make sure that dosages and side effects are carefully studied in children as well as adults. As we work to improve our public health system's ability to respond to bioterrorism, we must do a better job of addressing the specific needs of children and make sure they are fully accounted for in all local and state planning as well as in the research for appropriate treatments.

How to Cope with a Major Attack without Afraid?

This is probably the question from parents. We all want to do whatever we can to make sure our children are safe. But we certainly don't want them to have to live in constant fear. So how do we allow our children to lead normal, happy lives while being aware of, and prepared for, potential bioterrorist threats? That's a tough question. But it can be done.

A lot of the specific answers depend on the age of the child, and I'll deal those differences in just a bit. But there are certain things that apply to all children. First and foremost, spend more time with your children. Reassure them that you love them and will do everything in your power to keep them safe. You can't just think it. You have got to verbalize it and express it openly. How you do this will depend on how old your child is, although hugs and kisses are universally understood, from toddlers to teenagers.

Remember, like it or not, they'll take their cue from you. We all now that children are a lot ignore perceptive than we often give them credit for. They will pick up on any fear and anxiety you feel. So be mindful of what you say and do in front of your children.

If you are feeling over by anxiety, anger, or grief, that is probably not the best time to talk with your children. But there is no reason to try to hide it, either. Just tell them right up front that you are feeling upset right flow, so it's not a good time to talk. Let them know that you do want to talk to them, though and that you will in a while, once you feel calmer. This lets children know that these feelings, as overwhelming as they may seem at the time, will pass and that adults can handle them. That can be have reassuring.

Just make sure that you really do talk with them later. It is important that parents don't become so wrapped up in their own feelings that they ignore their children's needs.

Also, regardless of how old your children are, listen before you talk. The only way you'll know what's really bothering your children is to let them tell you. Resist the temptation to lecture or tell them everything you think they need to know. Ask them what they have heard from friends, teachers, on the TV or on the Internet.

Ask them what questions they have, and answer them as best you can in a direct and straightforward manner; with limited detail. Too much information can be overwhelming, especially for younger children. Often, they'll have only a couple of simple questions. That is fine.

Just kccp in mind that your children may not always tell you in so many words that they are scared or worried. Those feelings may come through in their expressions, the way they play, or even by angry outbursts. Watch for dramatic changes in your children's behaviour or mood, and if those changes persist, consider seeing a mental health professional.

One other thing to keep an eye on is your children's TV-viewing habits. Now, given the quality of television programming these days, that's good advice anyway for parents. But it is especially true when the airwaves are filled with around-the-clock news of terrorism or war.

If the news is on, watch it with them so you can talk about it and answer any questions they have. And limit their viewing if it seems excessive or if it appears to be increasing their stress and anxiety levels. Though it's harder to monitor; also discourage excessive surfing on the Web in search of terrorism reports.

During these uncertain times, one of the most important things you can do for your children is to maintain family routines and traditions. Celebrate holidays and birthdays with all the joy you can muster. Make sure your children eat a balanced diet and get enough sleep. And be sure to carry on the regular activities that have always been part of your lives, whether it's sports, music lessons, religious services, or scouting.

Kindergarten and Elementary-School Children

Young children will hear information about *bioterrorism* from all kinds of sources schoolmates, teachers, parents of friends, TV, even snippets of conversations overheard in a store. At this age, they are almost certainly going to misunderstand some of the information and have trouble separating exaggerations and rumors from fact.

They may think terrorists are in their school or believe that something bad happened that didn't. The main thing you need is patience and a willingness to listen. Find out what they have heard, and help them sort through what's real and what's not in a calm, gentle manner.

You may have to reassure the many times about the safety of your family and their school. Tell them the president and many other good people-firefighter police, military personnel and medical workers—are working hard to keep us all safe. They may ask you the same questions over and over, which every parent knows can be frustrating. Again, be patient. They may just need to hear the answers several times before they fully understand and accept them.

Children, particularly *preadolescent children*, may reenact traumatic events as they play (repeatedly wrecking planes and buildings, for example). Such play is often a way for children to attempt to master events that make them feel threatened and helpless. Reenactments—no matter how uncomfortable for adults—should be encouraged. Young children can often express their feelings and thoughts more fully through play activities than through talking. It may be helpful to provide a special time to paint, draw, or write about the events.

They may also ask what you consider shocking questions seemingly out of the blue. Remember, they are trying to process very confusing and frightening information, and there's no telling when something might pop into their minds. So try to respond in a calm, reassuring way, answering their questions simply and directly. It is important for them to know they can talk with you anytime.

Don't be surprised if you see some regression in young children, including whining, bed-wetting, or needing more help with getting dressed and eating. Be patient and understanding. And as part of your nightly routine, be sure to tuck them safely into bed and tell them you love them. Sleeping difficulties, especially nightmares, are common. Talk through nightmares with your children, and reassure them about their safety.

Middle-School Children

To children at this age, the grown-up world can seem scaly enough. You may see more aggression in some children, as they try to cover up fears and insecurity with tough talk. And you may see more emotion in others, as they cry and seek reassurance. Just as this is a time for adults to spend more time with one another it is important for adolescents to spend more time with their closest friends. Encourage this, even if it means you have to play taxi driver a little more often.

Some middle-school students find it helpful to keep a journal or express what they are feeling through drawing. If they want to share these with you, great. But don't force them to. It is also possible that middle-school children may talk more about death and dying. Or they may ask questions that include gruesome details or focus on death. This is natural, as they try to comprehend what's happening in the world. Respond calmly. If these comments become persistent, seek professional help.

The main thing parents of middle-school children need to do is be willing to listen. Understand that, on any given day, you may be the last person your child wants to talk to. It's nothing personal. Think back to when you were their age. How much did you talk to your parents? But your children will know if you are really interested in what's on their mind. Create a warm, understanding environment for them to share when they are ready.

High-School Children

Take time to discuss the numerous physical and emotional signs of stress. This will not only help them under stand their own reactions to events but also will teach them important life skills they will need as they prepare to enter the adult world.

Don't talk down to them or try to shelter them from what's going on. They want, and deserve, straight talk from you. For many, the events of September 11 and the months that followed will be for ever fixed in their memories.

Without downplaying the tragedy of all those who died, be sure to emphasize the way our country pulled together in a common cause. Talk about all of the people who sacrificed to help others and make a difference. One of the dangers is that, just as teenagers are about to enter the adult world, it can appear to be a frightening and uninviting place. We can't allow them to grow disillusioned. We owe them a better, brighter future. Talk with them about how world leaders could help reduce hate and violence, and encourage them to get involved in school religious or civic organizations.

Many high-school students are searching for, or have only Just begun to formulate their own ideas about, what they should believe in They may have idealistic concepts of right versus wrong peace versus war, etc. You may not think so, but as a parent you have a great deal of influence over how your children's beliefs develop. Discuss current events with them, allowing them to raise their own questions about what is right or wrong. As you begin to better understand some of the values your children have started to develop, you will be better able to help guide them.

2

AGENTS OF BIOLOGICAL DISEASES

For most of history, people have believed that the transmission of disease is a mysterious phenomenon controlled not by humans but by the *gods*, *witchcraft*, or *fate*. Collective disease, including the sudden epidemics that decimated cities and armies, was a frequent but misunderstood occurrence. Often relying on the ancient Greeks, Western scholars and physicians from the Middle Ages well into the nineteenth century held firm to the notion that "*miasma*" the stench of putrefaction, was the source of many epidemics and that changes in the weather or the planets increased the chances of outbreaks. One of the positive results of the belief in miasma was the instigation of campaigns to clear cities of garbage, open sewage, standing water, slums, and unhygienic slaughterhouses, measures that significantly reduced the risks of infectious diseases.

In both Europe and Asia, public health measures allowed urban economic growth. But whole societies remained vulnerable to devastating epidemics that were explained either fatalistically or by a science still struggling for proof. Toward the end of the nineteenth century, before any state biological weapons program, medical scientists discovered micro-organisms and made great strides toward understanding that a specific germ can cause a specific disease; that food, water, and personal contact communicate illness; that a pathogen can cycle through different species; and that insects and protozoans play a role in creating epidemics. Once these causal links were understood, humans could methodically control disease outbreaks. It became more possible to protect populations from the great assaults of plague, cholera, diphtheria, smallpox, influenza, and malaria that had swept across nations in previous centuries, hitting hardest in crowded urban centers and among the poor. Scientific knowledge by itself was no magic wand. Wars, forced migrations, famines, malnourishment, preexisting illness, and great poverty, especially in the colonial empires, remained the political preconditions of epidemics that science alone could not address then any more than it can now.

Together, stability of life, public health campaigns, and scientific knowledge about disease transmission aided human survival. By the 1920s, Western societies were rarely susceptible to sudden deadly outbreaks of the sort that threatened social order. Public health in cities had improved, water and food sources were monitored by the state, and vaccines and drug therapies were being invented as further protections. With most childhood diseases conquered, more people were living longer, a trend that has continued, and they were dying of the diseases of industrial society that afflict older people, such as cancer, heart disease, and stroke.

In the rest of the world, scientific knowledge alone without public health and freedom from war and poverty failed to prevent large epidemics. The enduring *dichotomy* between developed versus developing nations, or North versus South, remains marked by generally good health versus widespread, preventable epidemics. As Western nations were distancing themselves from the collective catastrophes of epidemics, some of their governments invented biological weapons as a means of achieving advantage in warfare. The German military during World War I made the first foray—against animals, not people.

It mounted an international sabotage campaign to kill packhorses and mules that were being shipped to the British and French from neutral nations, including the United States, Norway, Spain, and Romania, and from South American ports. Stevedores were paid to infect individual animals with anthrax or glanders and entire shiploads of animals were sickened and killed From the German perspective, these attacks violated no international norm. For years, though, this innovation cast suspicion on what further covert use of biological agents Germany might devise. This chapter begins with the establishment of *germ theory* and then proceeds to an account of the early days of *biological weapons*, when the French sought to best the Germans by *Integrating aerosols* from bombs with the new potential of air warfare. The tension between preparation for reprisal against an enemy nation and prohibitions against biological weapons characterize this early episode in biological weapons history. At the same time as the French were signing the 1925 Geneva Protocol, they were developing a biological warfare program to complement the one they had established for chemical weapons during World War I.

Little is known about how scientists first approached germ weapons. Included in this chapter, therefore, is a review of a remarkable early document, "*Bacterial Warfare*?' It was written in 1942 by two American biologists, Theodor Rosebury and Elvin A. Kabat, soon after the Japanese attack on Pearl Harbor, when scientists were joining up for the war effort. Classified at the time, the report was published openly in 1947. Rosebury and Kabat and others like them stood on the brink of an entirely new endeavor, on which they were willing to embark as patriots. The report's summary of the potentials of biological weapons and how to defend against them still resonates today, more than sixty years after it was written. The statement also expresses appropriate moral hesitation about the use of biological weapons against civilians, balanced against the deterrent effect such weapons might have on the enemy.

Epidemics

Throughout the recorded history of epidemics, ignorance of disease transmission has increased the risks of death and illness. Without knowing the source of an outbreak or the cause of its spread, large populations have been left defenseless. The great *cholera* pandemic that devastated Europe in the 1830s offers a catastrophic scenario that microbiology could later help prevent, if governments acted openly to educate public and maintained basic public health services. The outbreak began in India in 1817, and, in a second wave in 1826, it spread to Moscow and from there into Western Europe. No one then understood that a bacterium, the *Cholera vibrio*, causes the disease or that it is communicated largely through feces-contaminated drinking water. The disease reportedly killed 40 to 70 percent of those infected in a few days or even a few hours. Militarily enforced quarantines of infected communities caused riots and violence. Throughout the cholera epidemic, European doctors explained that it was caused by "*atmospheric conditions*" and miasma or noxious stench.

In the 1800s, physicians had difficulty distinguishing one epidemic disease from another. Fevers, rashes, loss of appetite, and aching joints were symptoms common to half a dozen likely "*plagues.*" One exception to this diagnostic confusion was smallpox, with its distinctive rash, for which a preventive inoculation was possible, although not everywhere accepted. For centuries itinerant folk doctors in China and India used the scabs from infected victims to inoculate others against the disease. This practice was communicated via the trade route to Constantinople, where, in 1720, Lady Mary Wortley Montagu, wife of the British ambassador, discovered variolation (as it is now called, from the variola virus) and brought it back to the British court. This innovation caused a theological and medical controversy about deliberate poisoning of the body. In 1796, investigating why farming people were often unscathed by smallpox, the English physician Edward Jenner conducted experiments on inoculation and then developed a serum from cows that could be injected to prevent smallpox infection.

The medical world soon became divided between those who accepted Jenner's method and those who doubted or feared it. Another hundred years would pass before immunization and vaccination were understood. Vaccinations figure strongly in *biological weapons* history, as they do in military and public health history, for the protective benefits they confer and for the anxieties they continued to provoke about risky side effects and bodily pollution. In the last half of the nineteenth century, European scientists approached the study of disease as a quest for the secret laws of nature, which persistence, inspiration, a good microscope, and a rudimentary laboratory for experiments might reveal. In 1858, French physician Louis Pasteur published his argument that specific germs cause disease, which he based on fermentation experiments. Then, setting out to disprove the theory that germs generate spontaneously in organic matter, Pasteur showed that germs are actually invisibly airborne. In 1876, Pasteur's illustrious rival, the German physician Robert Koch, gave rigorous proof to the germ theory by his experiments with *Bacillus anthracis*, the same anthrax-causing bacteria that became a preferred biological weapons agent in the next century. Koch used his pure culture technique to track the life cycle of the anthrax bacterium, from its dormant spore form through its germination to a rod-shaped growing form and back again to the spore. Koch went on to discover the tubercle bacillus that causes tuberculosis and the *Cholera vibrio*, and he set the laboratory standards for isolating and growing bacteria.

Meanwhile, Pasteur invented vaccines, which he named from the Latin word for cow, *vacca*, in deference to Jenner's work. His *anthrax* vaccine, announced a success in 1881, helped end livestock out breaks (epizootics) throughout the world. *Germ theory*, like Lady Montagu's smallpox inoculation and Jenner's serum, was resisted at first by the medical establishment. Those experimental scientists who understood its importance began using new basic laboratory techniques to culture disease agents and to investigate vectorborne diseases. Western microbiology, building on the remarkable French and German advances: became the foundation for all biological warfare programs in the next century. It was by no means obvious that microbes could be used as agents for weapons. As far as was known, they were not stable outside a host. Unlike chemical compounds, they seemed too fragile to use as bomb fill, which would subject them to the mechanical shock of discharge and high temperature. Both of these qualities, stability and hardiness, had to be scrutinized before biological weapons could be thought of as feasible.

French Biological Weapons Program

In 1919, at the request of the French government, Auguste Trillat, director of the Naval Chemical Research Laboratory, conducted an inspection of a German pharmaceutical plant, as part of the oversight instituted by the terms of the Treaty of Versaffles and the Inter-Allied Control Commission. At the plant, the director of the bacteriological laboratories apparently confided that German research on biological weapons was continuing. The French were well aware of Germany's wartime campaign against pack animals, and France itself may have engaged in similar sabotage, with undercover agents and prisoners of war using glanders or anthrax against German livestock. Trillat's report back to the French government raised alarm. The Germans had their fears as well, stimulated by rumors about Soviet bacteriological bombs. But their military experts were skeptical about biological warfare, advising that it might backfire on the attacker, and they focused instead on *chemical weapons*. In 1921 the French War Ministry concluded that France should begin a biological weapons program. To lead their program, the military turned to Trillat, who had already conducted experiments on the airborne transmission of bacteria.

Trillat was probably the first government scientist to calculate the potential military value of biological weapons and to play a role in implementing them. He believed that liquid cultures loaded into shells bombs could be detonated to form "*microbial clouds*" with great infective power. Furthermore, he convinced French authorities that the most efficient means of attack would be biological bombs dropped from aircraft.' The future use of airplanes for chemical bombing was simultaneously envisioned by others, among them the French military leader Marshal Foch, who wrote in 1921, "The carrying power of the aeroplane is increasing. Improvements are made almost daily enabling greater and greater weights to be carried. These developments introduce an entirely new method for the large-scale use of poison gas. By the use of bombs, which are becoming increasingly efficient and of greater capacity, not only have armies become more vulnerable, but the centres of population seated in the rear, and whole regions inhabited by civilians will be threatened."

Trillat had limited confidence in technology for protecting soldiers against germ weapons, for example, via gas masks; nor did he believe that vaccines would offer full protection. He saw biological weapons doing the most harm behind enemy lines—against reserve troops or against civilians in industries and cities, and against livestock, crops, and water supplies. To build retaliatory power as a defense, he advocated research on *anthrax*, *brucellosis*, *cholera*, *dysentery*, *plague*, and *typhoid* as potential weapons.

Making a *pathogenic aerosol* was a challenge. Like many biologists at the time, Trillat believed that microbes were fragile but could be sustained out side a host with moisture, the way droplets from coughing or sneezing spread germs. His experiments confirmed his theory that slurries of agents best kept their virulence; he therefore rejected the idea that pathogens might be dried and efficiently disseminated, which scientists later proved possible.

In discussions then about the drawbacks of germ weapons, scientists doubted that microbes could withstand explosive impact. Trillat, however, believed that the shock of an instantaneous explosion would disperse microbes without destroying them, so that an infective biological aerosol from bombs was feasible. Trillat also saw the air as a "*vast field of culture*" that, provided it was humid, would encourage the growth of microbes and allow the effective dispersal of germs over a city. His approach was more than poetic, for he had studied variable atmospheric conditions and their impact on the virulence of microbial aerosols.

How far the French went with their biological warfare plans remains unknown, owing to the disruptions when Germany occupied France in 1940 and to French secrecy both before and after the war. Pasteur's institutional legacy gave France a strong system of national and international biological research institutes, which, along with scientific training in its universities and medical schools, provided a decided advantage in exploring biological weapons in addition to research at the Central Naval Artillery Laboratory, aerosol and bomb trials were conducted at Le Bouchet National Explosives Plant, and animal experiments were conducted at the army Veterinary Research Laboratory in Paris. The substance of this research is difficult to evaluate except through a few publications by Trillat and his colleagues. It seems that little was known even to France's allies, although the British and French cooperated before the war on testing chemical weapons in Algeria. For a period between 1927 and 1934, the French biological warfare effort appeared to diminish, most likely owing to France's commitment to the Geneva Protocol. Then the French perception of a German threat intensified Hitler came to power, with particular fears about German air attacks.

French anxieties were spurred in part by allegations in the press that German spies had been using the Paris Metro to test the airborne dispersion of biological agents. French scientists, in reviewing German literature, also found an unsettling optimism about the feasibility of biological bombs. One German source was quoted as saying "Victorious will be the nation that knows how to find the most virulent bacillus and the most effective vaccine for defending itself against it." At a meeting of French military personnel in 1937, the argument was made that the Germans in war would be interested in a general conflagration and therefore would be ingenious in mounting a full-scale war "to assure the destruction of soldiers and of civilians behind front lines, no matter what their age or sex."

Between 1935 and 1940, the French military was again actively researching biological weapons, which was not the case in the United Kingdom, the United States, Germany, or any other Western power. By this time, logistical problems were coming into sharper focus. The rationale for choosing among potential pathogens still had to be decided. What was the target—humans, animals, crops, or water? If human, the condition of the host population mattered; this meant whether its immunity was already weakened by the conditions of war, bad weather (especially the cold), fatigue, or demoralization, and whether age, sickness, or race might affect vulnerability. What was the preferred route of infection? *Aerosols* were considered most efficient; bacteria might also be introduced by a cut or insect bite or by food or water contamination. The pathogen had to be virulent, that is, capable of rapidly causing serious disease. Could it sustain its virulence over time, through the process of fermentation and after being loaded into munitions and stored? Would the munitions efficiently generate aerosol without inactivating the pathogen? In addressing these and other questions, the French prefigured the approaches of scientists in the later state programs. So, too, did their first choices of possible agents, for the diseases *tularemia*, *brucellosis*, *meloidosis*, *plague*, and *anthrax*. They also considered toxins produced by bacteria, especially botulinum toxin (produced by the bacillus *Clostridium botulinum*), a favourite in the later state programs.

In the open French literature, arguments also appeared for civil defense against a possible biological warfare attack. One, from a military physician, advocated a new organization (DCM, *Defense Centre Microbes*) and proposed three tactics: (1) heightened detection of pathogens by testing air, dust, water, and food and by capturing and testing rats in cities and ports; (2) individual protection of citizens with face masks and vaccines, and (3) the spraying of prophylactic "*clouds*" of antiseptics in buildings and other inhabited spaces, an idea earlier proposed by Trillat.

In the spring of 1940, as German troops advanced, nearly all French documents at the various research stations were reportedly destroyed or hidden and the program evaporated. Despite British fears that French biologists at one of the Pasteur Institutes might cooperate with the Germans on biological warfare experiments, no evidence for collaboration surfaced.

While the French took seriously the perceived German biological war fare threats, American military experts reacted with skepticism. The defenses available to counteract germs seemed greater than any danger of biological attack. In 1933, for example, Maj. Leon Fox wrote an article in The *Military Surgeon* in which he reviewed the possibilities of biological warfare, proceeding from intestinal disease to respiratory illness to insect-transmitted outbreaks. In the first category, he trusted modern sanitary measures that reduced the chances of contaminated water, milk, and food. As for respiratory diseases, he put his faith in both natural immunity and the fact that the worst of these (to his mind, *influenza*, *pneumonia*, and *epidemic meningitis*) were contagious and would rebound as badly on the side using them as on an adversary. Fox made an exception for *anthrax* (which is not contagious from person to person), but he did not consider it a serious threat. As for plague and typhus, the insect borne diseases he considered most important, Fox pointed to the relative ease with which fleas, lice, and rodent reservoirs can be destroyed to contain any outbreaks. Furthermore, he believed that pragmatic consideration alone would determine whether biological or, for that matter, chemical weapons would be used in future wars. He was pessimistic that biological weapons would make the grade, noting that "at the present time practically insurmountable technical difficulties prevent the use of biologic agents as effective weapons of warfare."

Rosebury and Kabat on Biological Weapons

Within a decade, as war broke out, biologists in the United Kingdom and the United States took on the insurmountable technical difficulties as challenges. In 1942, Theodor Rosebury and Elvin Kabat, from Columbia University's College of Physicians and Surgeons, mapped the components of a comprehensive biological weapons program—from choosing biological weapon agents to framing offensive and defensive objectives. In a year's time, with assistance from the British, the United States started its biological warfare program. Rosebury joined it as a division chief and served to the end of the war; his junior colleague Kabat also contributed research to the program.

Some fifty pages long, the Rosebury-Kabat report begins with instructive comparisons between chemical and biological weapons. Like Trillat, whose work they knew, Rosebury and Kabat observed that both *chemical* and *biological weapons* agents, invisible, intangible, and diffused over large areas, might have psychologically frightening (what they termed "*insidious*") effects on civilians. In addition, chemicals and pathogens can equally poison the body and cause sickness and death. What, though, were the distinctive characteristics of biological agents? The authors noted six features:

1. Infective biological agents have *incubation periods*. That is, their effects are delayed and can take days to appear.
2. Disease agents can also be *contagious*. That is, they can spread from one human to another, and they can be spread by animals and by contaminated food and water.
3. Disease agents are characterized by infectivity, the frequency with which they can induce symptoms among individuals in a group.
4. Infectivity though, can vary according to individual immunity. Individuals differ in their susceptibility to contracting a given disease.
5. Infective agents, unlike chemical agents, are not manufactured. They are life forms that proliferate.
6. Without their characteristic mammal host, disease agents tend to be unstable and lose their virulence, although virulence can be technically enhanced.

Each of these features spoke to technical problems that Trillat had confronted and with which the British and Americans would also contend. Much about the disease agents selected for study was unknown. The actual impact of any intentional epidemic was theoretical and would remain so.

Rosebury and Kabat found Trillat's preoccupation with slurries and humid atmospheric conditions less interesting than the transmission of germs through the air. They turned to research indicating that bacteria sprayed into the air can be recovered from surfaces hours afterward with no loss of virulence; the agents of *pneumonia*, *cowpox*, *influenza*, *polio*, and *tuberculosis* had more stability than once believed. Along with reported experiments on stability, Rosebury and Kabat used the literature on laboratory accidents to inform their discussion. For instance, a mysterious outbreak of *brucellosis* at Michigan State University was traced to aerosols; dozens of laboratory mishaps involving tularemia showed the sturdiness of microbes dispersed over distance and left undisturbed over time. These two scientists knew more about drying techniques, such as freeze-drying (lyophilization), than Trillat had in earlier days. All this knowledge pointed to a potentially successful integration of pathogens into modern warfare.

What would be the best delivery system for a biological agent? "The *air plane*," Rosebury and Kabat concluded, "is clearly the most useful means for the dissemination of infective agents." But the authors refrained from endorsing bombs or shells. Despite Trillat's confidence, much was still unknown about the dispersal and survival of bacteria in explosions. A low-flying air plane might, they thought, directly assault ground troops or civilians with biological weapons using vectors such as infected fleas to spread plague, as the Japanese were accused of using in Manchuria. Airplanes might potentially pollute American water reservoirs. Furthermore, the dissemination of airborne agents needed experiments, for example, about which freeze-dried agents might best lend themselves to infective clouds.

From the outset, the authors were against any first use of *bacteriological weapons*. They believed instead in the investigation of biological weapons for defensive purposes and in their development for deterrence. In their report they cited the 1942 radio speech by British Prime Minister Winston Churchill declaring that, should the Germans resort to poison gas, the British would retaliate "on the largest possible scale far and wide against military objectives in Germany." *Rosebury* and *Kabat* responded, "The likelihood that bacterial warfare will be used against us will surely be increased if an enemy suspects that we are unprepared to meet it and to return blow for blow:

Starting anew, Rosebury and Kabat imagined the requirements of the entire biological weapons production scheme. If the military intended industrial-level production, they pointed out, it would have to build or convert plants comparable to or larger than commercial vaccine factories. Highly infective agents such as those for *plague* or *tularemia* would need an entire isolated building. If the military developed highly virulent or drug-resistant variants, it should be on the alert for the problems of disease transmission that accidents with unusual strains could cause.

Candidate Pathogens

Most of the Rosebury-Kabat report is devoted to studious review of seventy likely agents. The authors rejected thirty-seven and proposed thirty-three for detailed consideration. By comparing their list to contemporary biological warfare agent lists, we can see that most of the agents remain. Bacteria like anthrax and tularemia and a few viruses notably smallpox and yellow fever, are relatively constant candidates, although it should be noted that anthrax at this time in the US was still considered more as the cause of epizootics (animal outbreaks) than as a hazard to humans, among whom cases were rare. A range of hemorrhagic fever viruses represent additions from after World War II, for example, *Lassa fever*, *Marburg fever*, and *Ebola*.

The two scientists scrupulously evaluated the chosen microbial candidates, using ten criteria that ranged from the agent's availability (whether it could be easily cultivated) to the possibilities of specific immunization and effective therapy—defenses an enemy might easily have or acquire. Other important criteria concerned what percent of an exposed population might likely fall ill or die and how quickly, if the disease were contagious and about all its possible modes of transmission, and if a dispersed agent might contaminate the environment or pose a retroactive danger to friendly troops as they attempted to occupy an area.

The goals of total war determined the selection of potential agents, with no necessary distinction between causing intentional epidemics among civilians and destroying their food sources. Thus, the military purposes of biological weapons were envisioned primarily "for the disorganization of industrial areas behind the lines or of army centers and camps; for use as a part of a '*scorched earth*' policy, and against valuable animals, food plants and industrial crops." The siege of cities using biological agents and how cities might break such a siege were also briefly considered.

Some well-known disease threats failed to qualify. *Leprosy*, for example, had too long an incubation period. *Diphtheria*, understood as a children's disease, had too low an "*attack rate*" in adults, that is, too few victims would be badly infected. *Smallpox* and tetanus presented the problem of an enemy's being vaccinated or immune from previous exposure. The virus for *poliomyelitis* cultivated poorly and had low infectivity rates. *Tuberculosis* was a chronic disease with "low casualty effectiveness."

Botulinum toxin was first on the Rosebury-Kabat list. It was described as the most potent of all gastrointestinal poisons, by a wide margin, deadly in small quantities and with a short incubation period. Smallpox failed to make the candidate list; vaccination would make it ineffective. The list also excluded cholera and typhus, which later interested US and other state programs. On the list were influenza and Weil's disease (*leptospirosis*), a usually nonfatal waterborne bacterial infection, neither of which generated much subsequent military interest. Dysentery (*shigellosis*), a scourge in previous wars, ranked low on this list (and others) because normal sanitary precautions could defeat it. Rosebury and Kabat's concern that malaria might be introduced by the purposeful release of mosquitoes, an idea floated by the French, was not generally shared, although the United States did extensive defensive and offensive research on mosquito-borne diseases in general during the war and later.

Select Agents

The consensus about potential biological agents has remained high for decades. Rosebury and Kabat varied only a little in their judgments from later assessments. They ranked *Bacillus anthracis* in its dormant spore form as overall the most important agent, noting that it "surpassed by few microorganisms in infectivity for animals, and by none in host range." Most of the information about anthrax infection among humans came from industry. A worker in a woolen mill, for example, might contract either *inhalational anthrax* or *cutaneous anthrax*, through a skin abrasion or cut.

In areas of the world without reliable animal vaccination and veterinary inspection, infected meat could cause intestinal anthrax, also highly lethal. There was a theory, dating back to Pasteur, that abrasions in the respiratory tract aided anthrax infection and that inhalation anthrax itself, also called "*woolsorter's disease*;' was associated with large amounts of infected dust, a "*mechanical irritation factor*" that increased the chances of infection. Because there was no fully reliable therapy, anthrax was and remained too dangerous for human experiments to test such theories. Rosebury and Kabat understood that untreated inhalational anthrax was associated with high fatality rates. They believed its incubation period to be short, from a few hours to several days; although this was true for most victims, longer incubation periods for some people were later suggested by data from animal experiments and by the 1979 Sverdlovsk outbreak.

From a weapons perspective, the hardiness of anthrax spores was their outstanding characteristic. They lasted in the environment for years in cold and heat without losing virulence; in the laboratory, they could tolerate up to 100 degrees centigrade when dry. As for protection, the authors noted a problem: Allied troops would have to be vaccinated but no vaccination was prepared, except for animals. Even for animals, vaccinations had negative effects, with as much as one percent fatalities in herds. Well ahead of their times, Rosebury and Kabat called for the development of "a completely innocuous but effective [anthrax] vaccine." Decontamination of anthrax spores after an attack, they predicted, would present great difficulties.

Plague was ranked high as a disease candidate. In human history, *bubonic plague* (referring to the buboes or swellings in the armpits and groin) has caused devastating epidemics spread by infected fleas hosted by rats. If the infection in humans spread to the lungs, pneumonic plague, with higher fatality rates, sometimes followed from bubonic plague and spread through close person to person contact. The *plague bacillus* proved so stable in laboratory experiments that freeze-drying and dispersing it as an aerosol—to create *pneumonic plague*—seemed possible. "There is no reason to doubt that virulent plague bacilli could be disseminated by the airborne route, and that under conditions which can be rather clearly defined, a devastating epidemic could result."

High on the Rosebury-Kabat list, agents for tularemia and brucellosis later emerged as important in biological weapons programs. After the war, years of US human subjects research on the dose response for *tularemia* (also called *rabbit fever*) allowed its agents to become standardized for munitions fill, along with anthrax and yellow fever. *Brucellosis* has several recognized varieties that can infect animals (especially livestock) and humans. For humans, *Brucella melitensis* causes the most severe symptoms. The disease, with an incubation period anywhere between four days and four months, can last as long as three months with possible later recurrences. With a low death rate, estimated as less than 2 per cent, brucellosis would figure in later programs as a "humane" alternative to more deadly diseases. In addition, the virus for psittacosis (parrot fever) was judged potentially "one of the most useful agents of biological warfare" for its high airborne infectivity, though it is less fatal than anthrax or plague.

Creating effective aerosols for most of these and other pathogens was a problem that only animal experiments could solve, with controlled temperature, humidity, and wind direction and velocity. Rosebury and Kabat recommended that this research be conducted by universities and they pointed out that public health benefits might also accrue from such investigations of airborne infection.

Second in importance to airborne agents were vectorborne diseases communicated by fleas and lice, mosquitoes, ticks, sand flies, or other vectors harbouring parasites, whether bacteria, viruses, or protozoans. These possibilities, too, needed to be evaluated by experiments. The vectorborne diseases raised questions about which global areas might be attack targets. Mosquito-borne dengue fever, for example, characterized by relatively sudden onset (usually less than six days) with a slow convalescence, was identified with tropical climates. So, too, was yellow fever, also transmitted by mosquitoes and, although known in the Americas, of potential devastating consequence in Asia. Rosebury and Kabat suggested that some rickettsiae (the microorganisms in cells that can be transmitted by biting arthropods), like that for Q (query) fever, might be airborne. Tickborne bacteria causing *spotted fever* and *Q fever* were ranked high on infectivity, severity of symptoms, an extended convalescence, and possibly a high case mortality rate.

The two scientists more briefly discussed potential biological agents for causing animal and crop diseases. Among diseases that affect animals, anthrax, glanders, Rift Valley fever virus, and equine encephalitis could seriously affect both animals and humans. *Foot-and-mouth disease*, they noted, is highly infective for cattle, sheep, pigs, and other animals, but not for humans. Several viruses affect chickens and pigs. *Rinderpest virus*, a disease of cattle and buffalo, was noted for its severity; Canadian biological warfare scientists would later argue its threat to North American cattle and strongly recommend the development of a new vaccine. Bovine pleuropneumonia (*Mycoplasma mycoides mycoides*), eradicated from the Western hemisphere by 1895 and a danger to Asia, completed their list.

Rosebury and Kabat presented the possibilities of dozens of crop destroyers—bacteria, viruses, and fungi (most of them plant- and crop-specific), as well as insects like the boll weevil and the corn borer that might be launched against crops. Experiments, they again advised, would be needed to test the effectiveness of animal and plant pathogens and what technical defenses were possible. Plant pathogens proved the least favoured among biological weapons options. Once released, they might prove hard to check or eradicate.

Military Defenses

Defending soldiers against possible biological weapons attacks was part of the mandate of all biological warfare programs. Troops needed to be protected against infectious disease agents that the enemy might use, as well as against the pathogens used by their own side. Following the chemical weapons model, a simple, well-fitted mask could filter out pathogenic aerosols. In the event of an attack, soldiers could don their masks quickly, but they would somehow have to be alerted that they were under attack by an invisible aerosol, which would be difficult or impossible under battlefield conditions. For some but not all agents, prophylactic or therapeutic drugs, tested for use after exposure, could help protect the soldier.

Vaccinations or *antisera* promised round-the-clock protection, but not for most of the agents Rosebury and Kabat were ranking highly; furthermore, a large dose of an agent could over come such protection. In addition, some biological agents might, as the authors noted, be combined or changed in laboratory experiments to enhance virulence. If the enemy chose an unknown weapon that intelligence sources had failed to reveal, that surprise factor could overwhelm all defenses. No biological or chemical weapons were used during the war on any front. In fact, World War II proved a turning point in protecting soldiers in battle from disease and infection through technological innovation. Vaccines, sulfa drugs, the synthetic antimalarial drug atabrin and other derivatives, DDT (which protected against malaria and typhus), better techniques for the use of blood plasma and whole blood, and, later, the mass production of *penicillin* all contributed to better survival rates.

Civil Defense

Rosebury and Kabat did not flinch from the contradiction between biological weapons and the aims of public health. In public health, they observed, the main goal is to discover as much as possible about all the factors that spread disease, in order to break the causal chain.

In contrast, to develop disease as a weapon reverses this goal of public health. As Rosebury and Kabat described, "For the student of bacterial warfare the emphasis shifts fundamentally. He is concerned with the weak links only in order to strengthen them, or to discard the whole chain if they cannot be strengthened. He is much more interested in the inception of mass infections than he is in the details of their perpetuation, except, of course, that for purposes of defense against bacterial warfare, or for control of mass infections once they have begun, he must fall back on knowledge of the whole epidemiologic chain."

It followed logically that protection of civilians against biological weapons would be centered on public health reinforcement. Early detection was essential for the containment of a deliberate epidemic. Physicians, especially those in important industrial areas had to be well informed They needed to be alert to the use of potential agents know how to diagnose the diseases caused by select agents and be suspicious of unusual symptoms or circumstances in any disease outbreak.

In addition Rosebury and Kabat proposed three levels of civil preparedness, which were never fully implemented but bear comparison with cur rent American domestic preparedness plans. *The first level was emergency response* which would require amplification or mobilization of existing public and private health service facilities, in support of this level, every large military establishment and every large industrial area should maintain mobile "medical bacterial warfare units" that would cooperate with local health, police, and civilian defense agencies and have the power to enforce quarantines, evacuate unaffected groups from dangerous areas, and be involved in distributing face masks and making sure that the appropriate vaccines and prophylactic and therapeutic drugs were available. These units would also be charged with informing the public about individual sanitary and other precautions once the agent was identified. Air raid and fire wardens could serve as decontamination and agent identification squads.

As a second level of preparedness, they recommended permanent or semi-permanent measures for reducing the risks of respiratory diseases and of water and foodborne illness in workplaces and barracks. The methods they envisioned were those in use at the time: antibacterial sprays, ultraviolet lights, and chemical disinfection and sterilization.

As a third level of preparedness, Rosebury and Kabat addressed the opportunistic aspects of epidemics, advising broad public health measures to address social crowding, poor sanitation, and lack of vaccination and immunization in slums, already the breeding grounds of epidemic disease. "It would seem foolhardy in the extreme," they wrote, "to suggest that the possible consequences of a bacterial attack in such areas of congestion can be dismissed lightly." Emergency measures would suit the military and industry, but the civilian population would also need long-term protective policies to reduce the consequences of epidemics in general.

This advice was given in the context of the great resurgence in public health that marked the administration of President Franklin Roosevelt. New York City, where Rosebury and Kabat were located, had seen a tremendous growth in public health programs, for instance, in community health, child health, mosquito eradication, family nutrition, and dental care, and in new hospitals, local clinics, and laboratories. 'When advising emergency reinforcement of public health, the two scientists were referring to a strong infrastructure already in place, to which some but not great additions could be made. In the next fifty years, the American public health infrastructure was largely replaced by central hospitals, high-technology medicine, and curative care oriented to the individual patient as consumer.

After the war, Rosebury returned to medical science, writing about aerobiology and infectious diseases while also warning of the dangers of biological weapons. His 1949 book *Peace or Pestilence*? remains a classic that outlines the risks of biological weapons projects and asks that the public make itself aware of the gravity of the policy issues. Kabat continued as a consultant to the Chemical Corps until 1947. He became a distinguished immunologist, receiving the National Medal of Science from President George H. W. Bush in 1991. It recounts, the tension between wartime public health and biological weapons or "*public health in reverse*" dominated British government discussion in the years just preceding World War II. As the war became a reality, public health advocates went one way, promoting general strengthening of the British healthcare system. At the same time, other civilian biologists of high caliber took a distinctly different path and joined with the military to develop biological weapons. The British program was not the world's first venture in this area. France and then Japan preceded it. Rather, the UK effort was the world's first technical and organizational success, the innovation that moved biological weapons off the drawing boards and into the reality of munitions, industrial production, and simulated air strikes.

3

LABORATORY REQUIREMENTS

Like epidemiology, the *laboratory component* of the public health and medical response to bioterrorism is largely an assessment rather than an awareness tool. Laboratories work closely with epidemiologists to determine the nature of a disease outbreak, whether natural or deliberate. Laboratories will be critical in identifying the etiologic agent, cueing public health officials about the most effective response, and ensuring that physicians administer appropriate treatment regimens.

REQUIREMENTS

Sampling

For laboratories to be an effective assessment tool for responding to bioterrorism, laboratory technicians must be more aware of the *bioterrorism issue* and more capable than they are today of identifying potential *bioterrorism agents*. Unfortunately, physicians are generally reluctant to order more cultures given pressures from insurance companies and by *Health Maintenance Organizations* (HMO) to reduce the number of non-essential procedures. In this environment, physicians are unlikely to refer a culture sample to a clinical laboratory unless something unusual is suspected, thus placing a heavier burden on surveillance to detect an outbreak. Given that HMOs operate in a for-profit culture, it will be difficult to convince them to pay for additional laboratory analyses as a mechanism for detecting bioterrorism. Bringing HMOs into public-private partnerships to address the threat of bioterrorism is an important requirement.

Surveillance

Clinical laboratories, or laboratories that receive routine culture samples from physicians for confirmation of syndromic diagnoses, represent part of the front line in a *bioterrorism event* and are essential for effective utilization of laboratories for *surveillance* and early detection of an unusual outbreak of disease. A drastic increase in culture requests, the appearance of an unusual sample, or a cluster of samples with the same characteristics may indicate a bioterrorism event. Because physicians may not associate clinical flu-like symptoms with a *bioterrorist attack*, clinical laboratories receiving patient samples may be the first to detect a bioterrorist attack if they are aware of what they should be looking for and what to do if they suspect something unusual is occurring.

Diagnostics

Assuming that physicians take cultures from patients exhibiting *flu-like* or unusual symptoms, and assuming that laboratory technicians detect and report an unusual culture sample, forwarding a sample to a laboratory that can rapidly diagnose the etiologic agent is critical to initiating proper medical treatment and response procedures. The sample also will be an essential source of information for epidemiological investigations of the location of an attack and the people potentially exposed. Given the short time period between exposure to many bioterrorism agents and sickness and death, rapid diagnosis and treatment are absolutely essential to minimizing fatalities. Physicians will depend on laboratories to determine the agent used in a bioterrorist attack in order to treat patients effectively. Laboratories must be able to test for antimicrobial sensitivity and determine whether a particular antibiotic will be effective against the given agent.

Likewise, public health officials will need to know the nature of an outbreak in order to institute the appropriate public health measures to protect uninfected people. These measures will be drastically different for *anthrax* than for *smallpox*. Laboratories will also be important in determining whether more than one attack has occurred and how many agents are involved. Laboratories must also support law enforcement efforts not only by identifying the agent used in an attack, but by conducting microbial forensics to determine where the agent may have originated.

There is also a need to enhance laboratory capabilities to identify a wide-range of potential bioterrorism agents. While the CDC recognizes that the *bioterrorism threat* is broader than the five *bioterrorism agents* it has identified as having the greatest potential for causing mass casualties, it has only recently added 14 more bioterrorism agent protocols to the list of laboratory procedures. These protocols, however, have not been fully disseminated or integrated into most laboratory operating procedures.

Ultimately, more attention must be paid in the laboratory context to the potential for a *bioterrorist attack* with agents that have not traditionally been weaponized by militaries, particularly those that have been the source of recent disease outbreaks. *Salmonella*, *E. coli*, and *cryptosporidium* are examples of agents that may be used by terrorists, yet are unlikely to raise suspicion among laboratory technicians that a biological attack has occurred. They could assume, for example, that an outbreak caused by these agents is the result of unsanitary food service or handling conditions. As the Rajneeshee salmonella attack demonstrated, a link to bioterrorism may only be discovered later. Conducting advanced microbial analyses on agents like *salmonella* or *E. coli* could determine that rare or multiple strains are involved in an outbreak indicating the possibility that the outbreak was intentionally introduced. Building the laboratory capacity to routinely conduct advanced microbial analyses on common agents is not feasible, yet an unusual surge in culture requests for one of these agents should prompt laboratories to forward samples for more advanced analyses. A balance must be drawn between building laboratory capacity for identifying the most likely bioterrorism agents and less likely, but still possible, contingencies.

Surge Capacity

State and local laboratories possess a range of capabilities and perform different functions, from simple culturing to genetic sequencing. Some of these facilities are public and others are private; some are housed in hospitals and public health departments, and others are independent entities; some handle an enormous volume of cultures and others far fewer. For these reasons it is difficult to place the onus of building laboratory capacity on a single type of laboratory. Like other functions in the bioterrorism response system, providing a surge capacity is critical. Such a surge capacity requires laboratories to cooperate during a *bioterrorist event*. Having the ability to refer culture samples within a network of laboratories with varying capabilities provides the necessary technologies and surge capacity that will prevent backlogged culture requests during an event.

Personnel and Equipment

Conducting rapid diagnostics and producing surge capacity cannot be done without equipment and trained personnel. The number of personnel available and trained to conduct culture analyses of potential bioterrorism agents during an event will be critical, as will the amount of technical equipment to conduct identification analyses. For each positive confirmation of a clinical diagnosis that laboratories make during a bioterrorist event, epidemiologists will have another patient profile that can be used to characterize the nature and scope of the attack. This information will prove useful for guiding medical management decisions.

Eventually technological tools may provide rapid diagnostic capabilities in accurate hand-held assays that can be used during a bioterrorism event to triage patients and quickly decipher the "*worried well*" from the truly ill. Other technological tools that will aid laboratories include shared access to databases housing molecular sequencing of pathogen strains, microbial sensitivities, and areas of origin for use in defining both appropriate public health measures and law enforcement actions.

Training

Educating state and county laboratory personnel in using new techniques and equipment is also an important component of building laboratory capacity, especially given advances in molecular biology and biotechnology. Technology-based training is needed to familiarize employees with new technologies that will enable state and county laboratories to perform more advanced tests during a bioterrorism event. Using a set of "*unknowns*" may also be a helpful training method. Personnel must also be trained in agent-specific protocols for sampling, conducting laboratory procedures, transferring samples to better equipped laboratories, and chain of custody issues so as to get accurate results and maintain evidence for law enforcement. As training continues for laboratory personnel, bioterrorism issues should be included in training proficiency testing to gauge how well laboratory personnel understand the procedures that will be involved and roles that laboratories generally will play in responding to a bioterrorism event.

Communication and Information Sharing

The ability of laboratories to communicate and share information with other laboratories, as well as with public health departments, hospital staff and physicians, and other response entities so that all those involved in responding to a bioterrorism attack are well informed and acting on the same information will aid a more efficient response. The CDC is the principal hub of communications among laboratories. The Public Health Laboratory Information System (PHILS) is the information sharing system that is currently used by the majority of laboratories that communicate and send information to the CDC. The CDC should work with laboratories to standardize software and IT systems to facilitate the exchange of information and allow laboratories to communicate more effectively. This would allow for better integration and cooperation between laboratories and other public health and medical entities in monitoring infectious diseases and responding to bioterrorist attacks.

The CDC and the *Association of Public Health Laboratories* (APHL) have organized and are expanding the *Laboratory Response Network* (LRN), a series of laboratories that assist one another in the event of a bioterrorist attack. The LRN establishes cooperative arrangements between network laboratories with varying degrees of capabilities to provide the necessary diagnostic and support services. The LRN provides laboratories with a surge capacity to handle increases in specimen analysis requests that could result from a bioterrorism attack by allowing local laboratories to distribute the load among many laboratories in the network as well as by providing the necessary personnel, technical equipment, and diagnostic services that may not be available at lower level laboratories.

The LRN also utilizes the existing capabilities of laboratories to ensure that a wide-range of procedures and capabilities are available to perform diagnostics in the event of a bioterrorist attack. The LRN extends from clinical laboratories through state public health laboratories with more advanced capabilities to the most advanced laboratories at the federal level. Any suspicious samples identified by lower level laboratories will immediately be forwarded up the chain. In the case in which a specimen is thought to be related bioterrorism, arrangements have been made for immediately forwarding the sample to federal laboratories at the CDC with Biosafety Level 3 and 4 capabilities for working with the most dangerous pathogens like *smallpox*, or *viral hemorrhagic fevers*.

The CDC *Rapid Response and Advanced Technology* (RRAT) laboratory will initially receive an unusual sample and either identify the sample quickly or refer it to a divisional laboratory with agent-specific capabilities. These laboratories will confirm the results of lower laboratories, perform advanced microbial forensics, check the specimen against agents and strains formerly encountered, and bank the specimens for future reference. The specimen may also be referred from the CDC for evaluation by the laboratories at the U.S. Army Medical Research Institute for Infectious Diseases (USAMRIID) or

FBI laboratories, but that decision is made at the federal level and coordinated between DoD/USAMRIID, DHHS/CDC, and/or DoJ/FBI.

Once results are confirmed on the federal level they are reported back to state and local departments of health and laboratories so that appropriate public health measures can be taken. Essentially, the LRN provides a surge capacity and gives local level laboratories access to advanced capabilities, alleviating the need for costly upgrades to ensure laboratory capacity at the local level. All states now have some laboratory capacity to respond to a bioterrorism event, if not locally, then through the resources of the LRN.

All laboratories that are registered as part of the LRN are designated Level A, B, C, or D based on their capacity to support the diagnostic requirements of a bioterrorist attack. Level A laboratories include hospitals and other clinical laboratories, which receive culture requests from physicians to confirm clinical diagnoses. These laboratories are the front-line of laboratory capacity in the sense that they process the highest volume of culture requests and are the first to receive cultures from physicians. While Level A laboratories can confirm some specimens and rule out others, they cannot positively identify bioterrorism priority agents. Although Level A laboratories will likely be the first laboratories to receive a sample from a potential bioterrorism victim with *flu-like* symptoms, most have not been integrated into the LRN and have little understanding of how the network operates.

By definition, they have the least trained technicians and lowest technical capacity. In many cases Level A laboratory personnel are unaware of *bioterrorism issues* or potential bioterrorism agents. Technicians, for example, will dispose of culture samples containing *bacillus*, which is a common laboratory contaminant, without referring the sample to another laboratory for additional testing to rule out *bacillus anthracis*, the causative agent of *anthrax*. Level A personnel must be better informed about potential bioterrorism agents, how to report unusual cultures to the local public health department, and how to forward the culture to a higher level laboratory for additional testing. They also need to be sensitive to their role in surveillance, given that in the event of a bioterrorist attack, Level A laboratories will receive more culture requests of a particular type. Such surges must first be reported, and then be processed in a timely fashion to determine if something unusual is occurring. Level A laboratories must recognize the importance of these issues and have mechanisms in place for monitoring, reporting, and transferring samples in the event of a surge. If Level A laboratories do not recognize unusual agents or an initial surge in atypical cultures and refer these samples to Level B or C laboratories for further analysis, it may take several days for enough sick people to present before a public health official or physician recognizes that something unusual is happening. By this point, many people will be exhibiting symptoms and will be much more difficult to treat if treatment is possible at all.

Laboratory technicians, who fail to recognize the importance of, or dismiss unusual culture results, will delay detection of an outbreak and treatment of potential victims. Level B and C laboratories, are critical in detecting and shaping the response to unannounced bioterrorist attacks; their analyses of samples that may indicate suspect bioterrorism activities will serve as the basis for mobilizing law enforcement as well as federal assets. Because there are fewer Level B and C laboratories than Level A laboratories, they are well integrated into the LRN. Laboratory personnel in Level B and C laboratories have a relatively high awareness of how the LRN operates in the event of a bioterrorist attack and they are trained in new procedures using more advanced equipment.

Level B and C laboratories are usually county and state public health laboratories (SPHL) that have more highly trained personnel and better technical capabilities. They can identify etiologic agents, confirm clinical diagnoses, and conduct anti-microbial susceptibility testing so that the best possible medical treatment can be administered. The LRN has also established procedures and protocols for analyzing and transferring specimens, and communicating and sharing information among network laboratories. These protocols standardize the laboratory procedures in dealing with *bioterrorism agents* to ensure accurate results and preserve evidence for law enforcement. For example, in an announced bioterrorism event, mid-level laboratories will analyze the sample for law enforcement entities to determine if it is a biological agent. If it determines it is not, the FBI takes custody of the evidence and the sample goes no further.

Prior to establishing this as standard procedure, there was confusion over who should analyze the sample. Likewise, laboratory technicians were unfamiliar with priority bioterrorism agents or how to culture. The CDC and APHL have now established these protocols so that technicians have guidance during a bioterrorism event. The CDC and APHL have also held a series of training and education programs aimed at raising awareness of LRN laboratory personnel about bioterrorism issues and how laboratories fit into the public health and medical response to such an event.

Considerable progress has been made on the laboratory front over the past year. The CDC has aided laboratories through its grants process, providing funds to state and county laboratories to procure needed equipment and personnel for handling a bioterrorist event. Generally, bioterrorism grantee laboratories can handle initial response and surveillance requirements and conduct molecular analyses, but state and county laboratories that have not been recipients of grant money have little or no response capacity and no molecular methods capacity. Even without this capacity, all states have access to other laboratories in the LRN with sufficient response capacity, and molecular methods technologies.

Even with funds from CDC, however, state and local public health departments and laboratories are still constrained financially. Many laboratories are old and utilize outdated equipment. Some laboratories are also limited by their physical size and layout, which inhibits them from expanding to accommodate additional equipment and personnel. Some laboratories have rearranged parts of their facilities to make additions, but many others do not have this ability and cannot afford to expand. Molecular technologies are being increasingly used on the state and local level but state and local laboratories do not have a large staff that can exploit these techniques in the event of a surge in demand during a *bioterrorism event*. Therefore, such laboratories will largely rely on support from CDC in a crisis.

Communication and Information Sharing

Individual laboratories interface with LRN largely through a website. This simple arrangement is sufficient for obtaining bioterrorism training protocols and ordering and transmitting reagents, but it does not allow for substantial information sharing or actual interaction between individual laboratories. Laboratories lack the basic equipment necessary to communicate and share information with their partners in the LRN as well as with the local department of health and other public and private health institutions like hospitals, clinics, and primary care physicians.

This lack of basic information technology requires that personnel do potentially automated tasks by hand, which drains a laboratory's staffing resources. For example, laboratories and state public health offices keep databases on pathogen strains and their prevalence in local communities. Various databases, IT systems, and software applications are largely uncoordinated and result in duplication of infectious disease tracking by multiple organizations and institutions and inhibit the sharing of information across entities with different IT systems. This problem is probably most pronounced between private laboratories and public institutions and hinders the ability of Level A laboratories to communicate or provide surveillance data to the local department of health regarding disease outbreaks. Laboratory directors are concerned that these systems may drain staff resources by forcing staff members to constantly report information to different databases at the CDC, which often overlap.

Awareness and Training

The CDC has recognized a lack of knowledge about technological developments and how they can help in the laboratory setting. Therefore, CDC is providing trainers to work with staff in state health departments and laboratories to provide training in technologies like *pulsed field gel electrophoresis* (PFGE) and other molecular techniques. It provides training on technological advances dealing with food microbiology, fungal and viral infections, rabies, tuberculosis, and new and emerging pathogens. To conduct this training, the CDC has sponsored workshops around the country at which laboratory staff gain hands-on experience with the latest technologies. The CDC has also arranged 8 awareness-raising courses directed at Level A laboratory technicians and is funding state and county laboratory personnel to travel to the CDC to study advanced laboratory techniques.

The APHL has continued this training for members of the LRN and has developed web- based informational materials that can raise awareness of the bioterrorism issue. APHL has also developed the official protocols for analyzing the five bioterrorism priority agents: smallpox, plague, anthrax, tularemia, and botulinum toxin. Although these provide initial guidance for laboratory technicians, the CDC has recognized that these protocols need to be expanded to include additional agents and are in the process of doing so. In the near future laboratory technicians will have web-based access to an additional fourteen bioterrorism agent protocols.

Key Issues and Recommendations

Sampling

Shrinking health care budgets and HMO regulations have resulted in a reluctance by physicians to request unnecessary procedures, including requesting laboratory culture analyses. Given that many *bioterrorism agents* result in flu-like symptoms, physicians must be encouraged to take cultures and request laboratory analyses on a more routine basis to ensure that something unusual is not underway, especially if patients are presenting in unusual numbers or are exhibiting flu-like symptoms out of *flu season*.

Diagnostics

Laboratories are important for determining what agents are involved in a *bioterrorism attack* so that patients are treated appropriately. Given the short time period between exposure to many bioterrorism agents and sickness and death, rapid diagnosis will enable treatment to minimize fatalities. A number of steps should be considered

1. Level B and C laboratories should continue to upgrade their capabilities, including increasing the range of potential bioterrorism agents that they are capable of positively identifying, so as to reduce their dependence on the CDC laboratories.
2. State laboratories should also bolster their ability to test for microbial sensitivity and determine whether a particular medication will be effective against the given agent.
3. Technology-based training is needed to familiarize employees with new technologies that will enable state and county laboratories to perform more advanced tests and use these capabilities to support lower level laboratories if needed in a bioterrorism event.
4. The CDC's Rapid Response and Advanced Technology laboratory (RRAT) should continue to expand its technical capacity for rapidly diagnosing both critical agents as well as a broad range of potential bioterrorism agents that may be used in an attack. Because rapid diagnostics will be critical for early intervention, the RRAT should also expand its personnel to ensure that it will be able to support the State Public Health Laboratory during a bioterrorism event. Research and development needs to continue in the area of diagnostics tools that can relieve the burden placed on laboratories. In particular, hand held assays capable of giving a quick and accurate confirmation of patients at triage and point of delivery stations.

Awareness of a Wide-range of Bioterrorism Agents

Although the CDC has acknowledged a range of potential bioterrorism agents, the emphasis on a narrow set of bioterrorism agents has created a perception that a *bioterrorist attack* is most likely to involve agents that have traditionally been weaponized by militaries. This threat analysis has found otherwise, therefore, training for laboratory technicians is needed to expand their awareness of the range of potential bioterrorism agents.

Laboratory Surveillance

Laboratories have a critical role to play in *bioterrorism surveillance*. Level A laboratories could be the first to notice a surge in culture requests, unusually high disease prevalence, or the appearance of an unusual etiologic agent. Recognizing the implications of these indicators of a bioterrorist attack and communicating this data to the local department of health is critical to mobilizing a prompt public health and medical response. The LRN has been successful at creating a network of laboratories in each state with a range of capacities to bring to bear during a bioterrorist event, but it has been less successful at integrating clinical Level A laboratories that will be the first to receive culture samples in most cases. This is a weakness of the LRN, especially given that Level A laboratories have a role to play in surveillance and could be the first to notice a surge in culture requests or an unusual agent. Level A technicians as a result are largely unaware of bioterrorism issues. Integration into the LRN will help to raise the awareness of bioterrorism in Level A laboratories and what role they will play.

The LRN must expand its network of Level A laboratories and better integrate food, water, and veterinary laboratories to ensure that diagnostic capabilities for the range of bioterrorism agents are available. Laboratory technicians at all levels, particularly in Level A laboratories, should receive awareness training and proficiency screening for issues related to bioterrorism.

Surge Capacity

Laboratories must be prepared to handle an influx of culture requests in the event of a *bioterrorist attack*. While the LRN has done a good job of providing surge capacity for less-advanced diagnostic capabilities as well as access to a full-range of diagnostic capabilities through the network, the current surge capacity for meeting advanced microbial diagnostics requirements during a bioterrorist attack is questionable. The CDC should continue to provide funding through its grants process to build advanced laboratory capacity at the state level, which will reduce dependence on CDC for the provision of these services and will bolster bioterrorism assessment tools at the state level.

Communication and Information Sharing

Individual laboratories interface with LRN largely through a website, but lack the basic equipment necessary to provide automated surveillance data to the local department of health or communicate and share information with their partners in the LRN, hospitals, clinics, or primary care physicians. Laboratory communication and information infrastructure should be bolstered, particularly through greater automation.

Funding

Adequate funding for laboratories is central to building capacity and raising awareness of the bioterrorism issue. The equipment and techniques used in many laboratories are outdated and the facilities are limited on space. *Funding* is needed to renovate or redesign laboratory floor plans to ensure the best use of space and to accommodate additional equipment and personnel.

4

Medical Management

Prophylaxis

Prophylaxis in response to a bioterrorism incident means providing potential victims with the medicines necessary to prevent the onset of symptoms or the *vaccine* needed to prevent the spread of the disease through the population. In order for a mass prophylaxis program to be effective, it must provide the appropriate medicine to all individuals who are possibly infected rapidly enough to prevent disease onset and provide that medication over the extended period of time necessary to eliminate symptomatic individuals. Vaccines should be provided to the group of people likely to have come into contact with the agent. For *contagious diseases*, when available, vaccines must be distributed to certain segments of the population to prevent further transmission of the disease within an exposed population.

Planning for and executing a major rapid prophylaxis campaign is a challenge requiring development of pre-event plans based on possible contingencies. Included among the key decisions are:

1. Medicines or vaccines to be provided;
2. The segment of the population to be provided with prophylaxis;
3. The amount of medicine or vaccine needed to provide prophylaxis;
4. The distribution system most appropriate for the speed requirement and the decisions listed above; and
5. Any adjustments to these decisions as the event changes over time.

Prophylaxis should begin with identifying the agent used in the attack. Accurate identification will ensure that the most appropriate available medicine or vaccine is provided. In the case of *chemo-*

prophylaxis, there is a high degree of commonality between the ideal treatment options for many biological agents. Some degree of uncertainty in the type of agent used will not make prophylaxis impossible, but it does increase the risk that the prophylactic treatment used until the point of agent identification will be inappropriate or not ideal. This is especially true of vaccinations, given that vaccines are usually specific to agents or close families of agents. Identifying the amount of medicine required to apply prophylaxis successfully in response to a *bioterrorist attack* is a function of:

1. The amount of medicine needed per person per day to prevent onset; by
2. The estimated number of people who may have been infected or
3. The length of treatment regime to eliminate presentations of symptoms.

In the case of providing vaccinations, the amount necessary is determined by the number of people that must be vaccinated to prevent the further spread of the disease times the number of doses needed to achieve full immunity. The need to estimate the number of people infected reflects an important relationship between *prophylaxis* and *epidemiology*. Providing effective and efficient prophylaxis requires good epidemiological information—ideally to include an approximation of how an attack occurred operationally, the time frame for the attack, the method of agent dissemination, and the location of the release—to determine which people might have been infected. Without this information, everyone in the metropolitan area could be assumed to be potentially infected and should be prophylaxed.

For attacks in large urban centers, the blanket approach to prophylaxis might require millions of doses per day for periods as long as two months. With good information, however, epidemiologists can develop a profile of those individuals most at risk. These profiles provide a useful tool for sorting out those people who should receive prophylactic treatment and those who are either not infected or at low risk. Once the type and amount of medicines and vaccines have been identified and sources specified for providing the supplies, the final requirement for prophylaxis is a distribution system. Given the complexities associated with distributing medicines or vaccines to many people, it is also the requirement most likely not to be met.

Planning for and managing mass prophylaxis should be done at the local level. Local officials will know in detail the layout of their city, including the *population distribution*, road configuration and traffic patterns, an approximation of available local level resources, and available labour pools upon which they can draw. Local officials can also anticipate the types of problems that might be encountered during a mass prophylaxis distribution effort before they occur, thus facilitating development of plans that mitigate those problems.

In order to calibrate and adjust their prophylaxis plans, local officials should also have information on what resources will be available externally, including medicines and people, and how they will be deployed. Plans for *prophylaxis distribution* vary from city to city, but most candidate systems are derived from a few basic options. The first option is to canvass the population being provided prophylaxis to provide them directly with the medicine or vaccine.

Basically, a large team would move door-to-door to residences and employers to provide people directly with the appropriate prophylaxis. Some localities have even studied the potential applicability. Postal Service in providing this canvassing capability. Even if the Postal Service is used, the central problem with this type of system is the number of people necessary for distributing supplies. There is an inverse relationship between the size of the canvassing team and the speed of distribution. Fewer people could be utilized to cover the same area and number of people, but would take more time. When there is an urgent need to provide prophylaxis, time is not an available resource.

Another option is to establish *points of distribution* (PODs) throughout the area where the individuals targeted for prophylaxis reside. Of course, having thousands if not tens of thousands of people converging on several distribution points introduces serious traffic difficulties that must be addressed or else risk having large segments of the infected population not have the ability to reach the PODs in a timely manner. One solution to this problem is to increase the number of PODs. Higher numbers of PODs decreases the number of people served by each individual POD. Like the canvass approach, manpower requirements increase as the number of distribution points increases. Moreover, more PODs make the distribution system more complex, increasing the risk of a breakdown.

Some cities have taken a novel approach to the concept of increasing the *number of distribution* points by using existing distribution points for pharmaceuticals. This plan envisions providing existing pharmaceutical distribution points—drug stores, pharmacies, clinics, hospitals - with the appropriate medicine or vaccines and then directing people to their regular pharmaceutical provider for prophylaxis. This system allows people to go to the place with which they are most familiar, increases the number of distribution points to decrease traffic and confusion, and utilizes the labour pool available within the existing system.

Treatment

Medical treatment will be key in mitigating the effects of a *biological attack*. Although treatment requirements will differ according to the circumstances of the attack—large versus small numbers of casualties or the period of time over which patients present at care facilities—several elements will be essential to an effective treatment regime, regardless of the context: assessing resource needs, developing a standard of care, and determining appropriate treatment measures.

Resources needs

Resource needs fall into two broad categories: *material needs*, such as *medicine*, *equipment*, and *treatment facilities*, and *human needs*, such as medical staff and volunteers.

Material needs

Material needs will surface in three main areas: equipment, medicine, and space. To some extent, the amount of resources needed will be determined by the size and scope of an attack. In the case of a limited attack—for example, one resulting in 10 or 20 casualties—hospitals or care facilities will probably have adequate equipment and antibiotics to treat any incoming victims. However, if an attack overwhelms local health care capabilities, local authorities will have to call on state or federal resources to provide the additional resources needed for treatment.

Necessary *equipment* can range from personal protective supplies, such as latex gloves, to negative airflow rooms, respirators, and separate ventilation and waste collection systems. Blood, ventilators, and special treatment units are especially critical for effective treatment. As most hospitals have limited amounts of this equipment on-hand, if needed it will have to be procured from other local or regional hospitals and clinics. The amount of equipment needed must be assessed in terms of the projected number of casualties; logically, a greater number of expected casualties will require a larger volume of equipment. Because certain types of equipment will be stockpiled within the national stockpile, hospitals and primary care facilities will have to establish agreements with other medical centers, city response agencies, or private companies to provide excess supplies. Partnerships should be considered based on an organization's proximity to the primary care facility, the type of equipment it can provide, and the sophistication of its distribution methods.

Antibiotics are another critical treatment element whose availability must be assessed. Medication will be needed for treating those who are already ill in addition to preventing disease in those who may have been exposed. Because hospitals keep no more than a few weeks of medication on-hand, casualties mounting into the hundreds will quickly deplete available supplies. Several efforts have been considered to address the problem of sufficient medicines for treatment. The immediate supply problem could be resolved in part through mutual aid agreements between health care facilities; for example, an agreement established among several hospitals and clinics in the Washington, D.C., region provides for the *pooling of medication* in the event of an attack. Contingency plans of this type may be vital to saving lives, since time constraints limit support from state and federal agencies during the initial critical period of an outbreak. However, as with equipment procurement, partnerships should be established based on proximity and distribution concerns.

Finally, *adequate space* for treatment is an issue. How a region will assess available space will depend in large part on how it structures its local response plan; some areas have designated several hospitals as primary treatment centers, while others have placed one hospital or clinic at the forefront of a response. Whether the agent is contagious or not will also have an important affect on the community's space assessment; a non-contagious agent will allow patients to be treated in more than one facility, while a contagious agent will require *centralization of casualties*.

In the event of an attack resulting in a large number of victims, local care facilities may be confronted with more victims than they have bed space. Consequently, alternative treatment facilities may need to be established. Local response plans should contain specific contingencies anticipating such an outcome. *Potential care* facilities may range from older, non-operational hospitals, to community centers or elementary schools. If a non-medical setting, such as a gymnasium, is chosen as an alternative care site, it is important to determine during the planning stages whether the facility has the necessary infrastructure to sustain a full medical response. For example, during the recent TOPOFF exercises in Denver, patient treatment was moved to an abandoned factory when hospitals became overwhelmed. However, once there, workers discovered that the factory lacked both electricity and running water, making treatment impossible. Adequate planning must include requirements for the provision of resources found in hospitals or other care settings, from beds and ventilators to linen, food, and waste management services.

Human needs

Staffing will be a major concern for health facilities attempting to treat victims. In the event of a *bioterrorist incident*, staffing shortages may be severe, for many staff may fear transmission of the disease to themselves and their families, resulting in a high degree of *absenteeism*. Consequently, hospitals or primary care facilities must consider alternative personnel sources. As with medication and equipment, mutual aid agreements between local and regional health facilities may allow for doctors and nurses to be dispatched quickly to treatment sites. Similarly, local hospitals or primary care clinics may call upon EMS providers to participate in treatment activities, from administering care to discharging patients.

If local partnerships do not provide enough *staffing resources*, a variety of specialized teams have been established to improve the ability of the medical community to respond to incidents involving biological weapons release. However, for the most part, state and federal assistance teams have been developed independently of existing local emergency management structures. Consequently, the majority of the teams are unfamiliar with the response mechanisms of any given locality. Without a clear understanding of how these teams will be integrated with local resources, their utility is diminished.

Personnel requirements are exacerbated by the need to incorporate appropriate rest periods and stress management activities into a response plan. In fact, proper rest will be critical in successfully containing an outbreak. According to several officials involved in bioterrorism planning in Seattle, a major cause of the law enforcement failure during the WTO talks was a lack of rest periods for on-duty policeman. While health management of a bioterrorist event will require different skills and activities of doctors and nurses than of the public safety community, taking care to avoid staff fatigue and stress will be important to the ability of doctors and nurses to do their job.

Developing standard care procedures

Developing a *standardized approach* to patient management is needed both to avoid transmission of a contagious agent and to protect hospital staff from secondary infection. Because health professionals will be key caregivers in the event of an attack, the priorities of the health care facility or facilities responding to a biological weapons attack could be ranked as protection of the current staff and patients and then provision of the best possible medical care for infected patients presenting to the facility. Current hospital infection control procedures are appropriate for the level of risk involved with *aerosolized biological agents*. These procedures include both standard and transmission-based precautions.

Standard precautions are designed for use during the treatment of pre-hospital and hospital patients regardless of their diagnosis or presumed infection status, and are intended to reduce the risk of transmission of microorganisms from both recognized and unrecognized sources of infection. *Transmission-based procedures*, which are disease-specific, are designed only for the care of those patients known or suspected to be infected with transmissible pathogens. However, as most diseases associated with bioterrorism are not easily transmitted from person to person, these procedures will not be as critical, although certain contingencies require awareness of them. Those entities involved in treating casualties will need a working knowledge of isolation and infection control measures, and these measures should be standardized within any treatment facility.

Determining appropriate treatment measures

After the onset of illness, only general supportive care and specific medical treatment are left to health care providers. *Appropriate medical care* may involve life support measures, such as mechanical ventilation, as well as administration of antidotes. *Supportive therapy* may include attention to skin lesions, supplementary oxygen, pulmonary toilet, and treatment of complicating infections. Treatment will depend in large part on the characteristics of the disease, as well as on whether the agent is contagious or non-contagious. For a *contagious agent*, treatment will have to include the establishment of isolation wards and negative airflow rooms, and doctors and nurses will have to be outfitted in full Personal Protection Equipment.

However, these requirements will not be necessary for a non-contagious agent. Throughout the treatment process, attention must be given to symptoms that indicate early life-threatening deterioration. Medical personnel involved in treatment must also consider methods to discharge patients once they have improved, as well as methods to dispatch those who have died to mortuary facilities. While most hospitals and clinics implement standard discharge procedures during regular operation, the ratio of victims to staff during a *bioterrorism event* will largely determine discharge thresholds. If the ratio of medical personnel to patients is high, victims may be monitored until symptoms disappear entirely, and long hospital or clinic stays may be instituted. However, if the ratio is low, doctors and nurses may not be able to provide care to those who are not desperately ill, and a simple improvement in symptoms, or passing of a crisis point, may be enough for doctors to decide that the patient can continue with home-based care.

It is important to note that if the agent is contagious, patients will have to be kept isolated until transmission is no longer possible. Depending on the size of the treatment facility and the number of resources available, patients who are no longer critically ill may be transferred to a second facility to await final discharge. Because most doctors and nurses will be busy administering to the sick, medical students, residents, or EMS providers may need to be charged with collecting patient information and prescribing long-term prophylactic care or therapy to those being discharged. Similarly, students or EMS providers will need to coordinate the transfer of any fatalities to mortuaries or funeral homes. Fatalities will have to be carefully tagged and tracked.

Triage

It is estimated that when the Aum Shinrykyo attacked the Tokyo subway with sarin in 1995, three out of four people who sought medical care did not require treatment. Since that attack, those psychosomatic patients who seek unneeded care during a WMD terrorist attack have been termed the "*worried well*". In a *bioterrorism event*, in which symptoms take days to weeks to appear, the phenomenon of the worried well threatens to overwhelm the medical system with people seeking treatment. Given that medical resources are limited, triage will act as a critical part of the medical management system by sorting patients who truly need treatment from those who do not. If triage is performed effectively, it can substantially reduce the stress placed on the medical system. In the early stages of a bioterrorism event when a handful of patients begin presenting, hospital staff, alternative care providers, and primary care physicians who receive patients will largely perform triage.

In the very early stages of an outbreak of an unusual disease, some suspicion may exist that exposure to a bioterrorism agent has occurred, but there will be little certainty. As a result, officials will be reluctant to initiate mass prophylaxis. But as larger numbers of people begin to present for treatment and it is determined that many people were potentially exposed, or if the extent of the exposure is unknown, a decision will have to be made to begin mass prophylaxis to prevent more people from becoming ill. Effective triage of potential victims is essential in reducing the stress on care providers as it will direct victims in need of treatment to facilities that can provide such services and will direct all other potential victims to points where they can receive prophylactic or self-treatment regimens.

In such situations it may be necessary to distribute prophylaxis to large numbers of people in a short period of time to prevent them from becoming symptomatic. The local medical resources for treating and prophylaxing victims are limited. Triaging patients therefore is essential to efficient use and management of these resources. Once a decision to begin massive prophylaxis has been made, a large-scale mobilization of local, state, and federal resources will be needed. The task of deciding which patients should receive initial treatment at hospitals and which patients should receive initial prophylactic resources could be complicated. Limits in *prophylactic supplies* may raise ethical issues about who receives priority consideration.

Many response plans call for administering pre-incident prophylaxis to key personnel and decision makers, including medical and public safety personnel, to ensure that they remain in good health to treat potential victims as well as to political leaders who will ultimately be in charge of maintaining command and control of the situation. Large numbers of "*worried well*", however, are also likely to present themselves for treatment despite having no need for it, raising difficult decisions. This challenge could arise especially in the early stages of an event when local supplies of medicine, vaccines, and so on are sparse.

Although triage is wrought with operational and ethical dilemmas and will differ according to the circumstances of the attack (e.g., large versus small numbers of casualties, or the period of time over which patients present themselves) as well as local resources and organization, the following activities will have to take place regardless:

1. Establishing triage points;
2. Providing information to the public about triage alternatives;
3. Information collection and patient evaluation at triage points;
4. Supplying triage points with prophylaxis medication and staff;

5. Sorting of patients on the basis of—(i) Direction to self-treatment regimens; (ii) Direction to treatment area; (iii) Determine that treatment is unnecessary;
6. Follow-on mechanisms to monitor patient progress.

Triage Points

Establishing *triage points* for the appropriate treatment regime will reduce the number of potentially exposed victims who go to already stressed hospitals, alternative care providers, or physicians offices. Directing potentially exposed persons to triage points where they will be evaluated and treated accordingly is a key part of the necessary public information strategy. There should be many, geographically dispersed triage points to prevent a large convergence of people and vehicles on one or a few areas. In planning the location of triage points, consideration should be given to how well these locations will be able to receive large numbers of people as well as how population movements may complicate the delivery of supplies. Good locations for triage points might include areas that are easily accessible by major roads or highways or rail transportation to reduce vehicle congestion in the area.

Triage points might serve not only as places where potential victims are sorted by their treatment requirements, but also as *Points of Distribution* (POD) areas for delivering prophylactic medication to exposed persons who are not yet symptomatic. Once a determination has been made that a patient needs prophylactic treatment they should be provided with that treatment as well as directions for self-administering the medication and monitoring their health status. The best way to provide these services is to set up a station that is separate from the "*evaluation station*" whereby large numbers of exposed people can receive instructions on how to administer their prophylactic treatment and where questions can be answered and fears allayed. Triage points will also have to be equipped with transportation services for people in need of more elaborate medical treatment

In many cases, especially for a large-scale attack, local hospitals, alternative care providers, and physicians offices may be flooded with potential victims who do not follow public information guidelines but rather present at the place where they normally receive care. Under these circumstances each facility will need to establish triage points someplace on the grounds of the facility to sort out which patients should be admitted into the facility for medium to long-term treatment and monitoring, and which patients can be administered self-treatment regimens without entering and overwhelming the facility where physical space is scarce.

Providing Information to the Public

A good public information strategy can help reduce the stress on the medical management system by providing the public with details on the time and location of the attack and advising them on where to receive appropriate treatment. Much of the *challenge of triage* will be getting potential victims to the right treatment area. In the case of a non-contagious release, the public will need to know that prophylactic measures can treat potentially exposed persons and where they can receive this treatment. Providing the public with information on time and place of the attack, and informing people that only those who can place themselves in the exposed area during a short time window need to seek treatment, can act as a self-selection triage tool. Public communications must also provide information on what symptoms may indicate exposure, where symptomatic patients should seek treatment, and where potentially exposed victims who are not symptomatic should go for evaluation and receipt of prophylactic medication. This information will be important for keeping hospitals from being overwhelmed. Meanwhile, in a contagious event, information messages should seek to convince potentially exposed persons to impose a self-quarantine by staying in their homes as the best way to contain the spread of the disease.

Information Collection and Patient Evaluation

Triage points will need to be staffed by personnel who can ask a potential victim a series of demographic and symptomatic questions to collect additional data that may prove useful to epidemiologists and law enforcement agents. Whenever possible, these interviews should be conducted by trained medical personnel who can make a decision about the appropriate treatment regimen. But in a bioterrorism scenario, numbers of real or potential victims seeking prophylaxis may appear, and there may not be enough trained medical professionals to staff both triage points and care facilities. Under these circumstances, a doctor cannot see every patient, but only those who, after initial evaluation, are determined to be at high risk of exposure. Initial evaluations will have to be conducted by public safety personnel, medical residents, or other available emergency response personnel. This determination will be made on how closely a patient profile fits the case definition constructed by epidemiologists and medical personnel, and on answers to a standardized series of questions that collect information on where the patient was at the day and time of the attack (if the approximate day and time of the attack is known) and what symptoms they are exhibiting. Based on this information, an initial determination can be made as to whether or not a patient was at a high risk of exposure and symptomatic, and therefore should be referred to medical personnel for further evaluation. In essence, if medical personnel are completely overwhelmed in a *bioterrorism event*, the collection of patient information acts as an initial filter so that only patients who truly need to see medical personnel actually do. The liberties that might be taken with this initial filtering decision will largely depend on the number of patients and the available medical resources and staff.

Staffing Triage Areas

Hospital staff, alternative care providers, and primary care physicians who may begin to receive the first victims of a *bioterrorism attack* will need to be notified that patients fitting the epidemiological case definition should be referred to a designated care facility where specialists can provide a more conclusive diagnosis. This initial triaging of patients by referral will allow epidemiologists to gather information on the location and time of the attack, which will be critical to making the decision about beginning large-scale prophylaxis.

Triage will take place at different places and the number of people who need to be triaged will not necessarily depend on the scale of a bioterrorism incident as many "*worried well*" will seek treatment. The triage for small numbers of exposed people requires little more than notifying care providers that they should refer patients fitting the case definition to a designated facility. As the number of potentially exposed individuals grows, triage requirements will increase and could necessitate massive numbers of personnel to interview, diagnose, and sort patients by treatment needs. Hospitals will have to set up a triage station on the hospital grounds to separate patients in need of immediate admittance from asymptomatic patients who require prophylaxis. Much of the sorting of patients must be done outside of the hospital facility to ensure that its interior can maintain normal operations in the face of increased treatment demands.

Points of distribution may also be set up to relieve the triage and prophylaxis burden placed on hospitals. When appropriate, POD stations can give symptomatic patients a certified referral for admittance to treatment facilities for treatment and monitoring. Once a decision to begin *mass prophylaxis* has been made, doctors, nurses, or other health technicians who are, by nature of their profession, more familiar with clinical recognition and treatment of disease, should make triage decisions whenever possible. However, in the case of an extremely large number of potentially exposed people requiring prophylaxis, some initial filtering of patients may need to be done by medical residents, premed students, veterinarians, dentists, or by the fire department or other representatives of the public safety community to weed out patients who were either not at the time and place of the attack if it is known, or who do not fit the case definition. As the number of people requiring prophylaxis or extended treatment increases, non-medical personnel are likely to take over much of the triage responsibilities as medical personnel are needed at treatment facilities.

Logistics and Distribution

A significant challenge in treating or providing prophylaxis to victims of a *bioterrorist attack* will be ensuring that response entities have the necessary equipment, medical supplies, and personnel. Achieving continuity of supplies in a crisis environment will be a daunting task. Putting into place a plan for distributing supplies before an incident occurs will be necessary for equipping response personnel to do their job in a time of crisis.

The starting point for any distribution system is a needs assessment to determine what resources—antibiotics, occupied and available hospital beds, ventilators, personnel, etc.—are currently available at the local and state levels. Such information can then be collated into a baseline of medical resource data. In the event of a bioterrorist attack, knowledge of where these resources are housed will be important for making decisions about re-supply efforts. As resources outside of the immediate area of the attack—nearby locales, state, or federal—are brought to bear, officials will need to rely on baseline resource data to determine the best distribution of these resources. The availability of this information prior to a bioterrorism incident will give local planners an indication of how state and federal medical assets can be integrated with local assets to have the maximum impact. This data will also be important for identifying and transferring resources from facilities with normal patient admissions to a facility that is stressed by an influx of patients as well as for informing response personnel about which facilities have resources available.

Some mechanism to track locally available resources must be developed. Facilities in the immediate area of attack will bear the brunt of treatment and prophylaxis activities and thus require much of the available medical resources. For this reason, initial decisions about where medical supplies should be distributed will be straightforward. But as medical supplies begin to dwindle and more potentially exposed victims present for prophylaxis and treatment, additional resources will have to be identified to meet the demand. Some of this demand can be met by transferring resources from care facilities that are not receiving large numbers of patients. But to do so requires knowledge of where available resources are housed and the ability to transfer those resources to care facilities that need them most.

While facilities receiving the bulk of the patient load will need a constant supply of medical resources throughout the crisis, as the medical treatment challenge grows with time, many care facilities outside of the immediate area surrounding the release site will also be receiving victims and PODs may need to be set up to *triage* and *prophylax*. Understanding the rate at which supplies are being utilized, as well as the re-supply needs of secondary facilities and PODs, requires some means for tracking what resources are being used to ensure that some areas are not oversupplied while others are undersupplied. It will also be necessary to track supply levels at central distribution points that are receiving state resources or CDC push packages to determine what additional outside resources may still be necessary.

Having identified the amount needed to provide treatment or conduct prophylaxis, the next key question is identifying sources of medications and vaccines. A number of potential sources of medicines and vaccines exist at the local level including supplies already distributed to area medical care providers. Hospitals, clinics, and doctors' offices immediately surrounding the area of the attack may possess some excess capacity not needed to treat both normal patients and bioterrorism victims. Extended reliance on local supplies is problematic, however, for a number of reasons. Some vaccines and antibiotics, like *ciprofloxacin* or *anthrax vaccine*, are infrequently or never stocked by hospitals, perhaps because they are not useful in regular practice or because of high cost. Even for those medicines frequently used, it is unlikely that local care facilities will be able to provide anything but minimal amounts due to just-in-time medical supply systems, the use of hospitals' and clinics' supplies to treat *symptomatic individuals*, and the sheer amount of medicines that might be necessary to prophylax very large numbers of people.

One strategy for dealing with this problem is to establish mutual aid networks with care facilities in surrounding counties and cities.

Excess resources within the network could be gathered and brought to the area of the attack. Such a network would have to include an extremely large number of hospitals and care providers given the general lack of excess supplies at most hospitals and the amount of medicines that may be needed. In addition, the process of gathering and moving resources within the network will be complex and time-consuming. Another local asset that could be available is pharmaceutical warehouses or production facilities, but the number of facilities producing the exact antibiotic or vaccine necessary is likely to be small, and the chance of having those facilities located near the area of attack is very slim.

An additional option is to develop dedicated stockpiles of medicines and vaccines most likely to be needed for prophylaxis in response to bioterrorism events. Determining which medicines and vaccines and how much should be stockpiled is a challenge due to the variety of potential attacks. Working with state and local partners, the federal government must develop national stockpiles of appropriate pharmaceuticals for use in both prophylaxis and treatment. These stockpiles should be based on rough estimations of the amounts of antibiotics and vaccines necessary to meet the needs identified by answering the previous questions— which medicines might be necessary based on the agent used, how many might require prophylactic treatment, and how many doses would be needed per person to complete the recommended treatment regime. The types of medicines and vaccines stockpiled should be based upon a sound threat assessment that identifies the most likely agents to be utilized and should capitalize upon the degree of commonalty among ideal treatment modalities.

This challenge is especially *daunting* given the high degree of specificity associated with vaccines. Developing, stockpiling, and maintaining a standing supply of a specific vaccine like smallpox vaccine will be expensive and is useful only if smallpox is deployed in an attack. Given the difficulties associated with acquisition, manipulation, and dissemination, the use of smallpox by a non-state actor must be deemed unlikely. But it is possible, and it would be an event of potentially enormous consequences. Given these consequences, it may make sense to develop some smallpox vaccine as a hedge against this possibility. The amount produced will depend on the amount necessary to stop the spread of a *smallpox* outbreak, the costs associated with production and maintenance, and how that cost compares with the resources necessary to meet other requirements.

Having established baseline resource data at the local level, planning should take into account how state and federal assets can be integrated into the local environment. Statewide medical resources could be mobilized in an emergency, and some states like Illinois have pharmaceutical caches that can be utilized in emergency situations. The CDC's National Pharmaceutical Stockpile "*push packages*" need to be factored into local planning based on local needs assessments. The challenge to fully utilizing the CDC's pharmaceutical stockpile is logistical. These resources will be delivered to local airports and handed off to local officials for distribution. In order to distribute this amount of prophylaxis, mechanisms for distributing medical supplies should be identified well in advance of a bioterrorist attack. Otherwise, the push packages will not be used to their fullest capacity.

Once a local airport receives supplemental medical supplies from the CDC *Pharmaceutical Stockpile* (state caches should be shipped directly to PODs via rail or truck), they must be effectively distributed to the appropriate facilities and/or treatment areas via local transportation mechanisms. In order to establish a reliable distribution system, cooperative arrangements need to be set up to use all available distribution systems, public and private, to distribute supplies in the event of a *bioterrorist attack*. Possible medical supply distribution mechanisms that could be used in the event of a bioterrorist attack include:

1. Pharmaceutical supply houses;
2. Commercial shipping companies;
3. U.S. Postal Service;
4. National Guard assets;
5. Department of Defense local assets;
6. Public transportation;
7. Private transportation companies;
8. Rental companies.

At the local level, pharmaceutical supply houses ship medical supplies to care providers and pharmacies on a daily basis, maintain warehouses for storing supplies, have employees and equipment for loading and moving them, as well as established supply routes and trucks for distribution. In some areas local response planners are talking with pharmaceutical supply houses about creating cooperative arrangements to support distribution requirements during a bioterrorism event. This would include using their loading docks, personnel, and transportation vehicles to distribute warehoused medical supplies as well as supplies contributed by states or federal entities. Commercial shipping companies or the U.S. Postal Service also maintain crews and trucks for distributing goods that may be utilized during an event. Public safety and public works or local volunteers may also be employed to drive rental trucks or public and private transportation vehicles such as buses or taxi cabs that can be loaded with medical supplies. State level transportation vehicles or assets operated by the National Guard, or even local DoD transportation vehicles, could also be utilized to ship additional medical supplies to care facilities if local distribution mechanisms become overwhelmed.

In any case, identifying the available distribution mechanism, putting into place cooperative agreements to utilize them during a bioterrorism event, and having a pre-incident distribution plan in place can help to ensure a smooth transition of outside medical supplies to local officials. Regardless of the type of system, large labour pools could be needed to provide treatment and distribute prophylaxis to the population. Where might local officials obtain the manpower needed to execute their plans? Increasing the number of people providing prophylaxis at the POD will increase the rate of service, but other people must track individuals who have received medicine or vaccine. Additional staff will be necessary to support those directly providing the prophylaxis. People are going to be needed to provide medical information and advice. Others will be required for the simple but critical task of opening supplies from their packing and moving them to the front line providers.

People will be needed to track receipt and distribution of medicines at the individual POD. This is necessary to determine the rate of distribution and the point at which a call for additional supplies must be made. Public safety personnel will also be needed to provide security for medical supplies located at the PODs. Motivated by profit or fear of not being able to reach supplies quickly enough, people may attempt to steal materials from distribution points. A small number of conglomerated distribution points could be guarded by smaller numbers of people, but only with a reduction in the speed of distribution. With less supply located at the PODs, a large number of small distribution points reduces the impact of having the supply at a distribution point stolen. The more extensive the distribution system, the more people will be necessary to serve security functions.

At the local level, the public safety community represents one potential resource for this task. Police, however, would probably be heavily involved with security and maintaining public order. Likewise, paramedics and emergency medical service personnel will have their hands full responding to 911 calls and transporting sick individuals. One potentially untapped source could be the fire department, although *firefighters* may also be involved with security and responding to acts of vandalism and malfeasance. State and federal assets might also be useful in support of local prophylaxis efforts. National Guardsmen en masse could be trained and deployed to support prophylaxis efforts. Regular Department of Defense personnel, especially medically trained personnel like nurses and medical technicians, could also be useful. Like the national stockpiles, time will be required to move and then integrate available federal assets into prophylaxis systems. In fact, it may take days longer for a sufficient number of federal personnel to arrive than the push packages.

Another option is to organize a corps of reserve medical personnel from retired or former doctors, nurses, physician's assistants, veterinarians, and dentists. Such a corps could be created by asking people to join voluntarily and have their names included within a national database. During a bioterrorism crisis, members of this corps could be called up from within the area affected by the attack.

Where Are We?

The Office of Emergency Preparedness, Department of Health and Human Services (OEP/HHS) maintains the National Disaster Medical System (NDMS). The NDMS is comprised of two main elements. The first involves voluntary agreements between the NDMS and hospitals in major metropolitan areas. Member hospitals agree to commit a number of their acute care beds, subject to availability, to patients received through NDMS. The federal government pays the costs of care provided to NDMS patients received by the participating hospitals. Because this is a voluntary program, member hospitals can provide more or fewer beds than committed in the agreement. The second element of the NDMS is comprised of a number of Disaster Medical Assistance Teams that could provide both additional personnel and supplies for treatment and prophylaxis. Currently, the combined total membership in DMAT teams is over 5,000 individuals.

According to existing plans, the Office of Emergency Preparedness of the Public Health Service (OEP/HHS) would coordinate federal assistance in the area of medical treatment. Other agencies, specifically the Department of Defense and the CDC, would also provide information and services. CDC, for example, is to provide its push packages within 12 hours of their release by the Director, although in a best case situation, an additional 24 hours would be needed to breakdown, catalogue, and package the contents and then transport them to area points of distribution. CDC could also establish an epidemiology command center within 24 hours. With these timelines in mind, it is imperative for local health facilities to have their own contingency plans.

In that regard, the OEP is also responsible for coordinating the creation of Metropolitan Medical Response Systems (MMRS), a process in which over 70 cities have been asked to organize local medical capabilities into a response system for incidents of CBRN terrorism. Recognizing that health and medical responses to bioterrorism are different than for chemical or radiological incidents, within the last year, OEP has asked those cities that completed their original MMRS response plans to develop further response plans specifically geared for bioterrorism incidents. In order to fulfill their contracts, localities must develop detailed plans for how they would conduct a rapid and massive prophylaxis program. While the *bioterrorism* track is relatively recent, approximately 10 of the largest MMRS cities have been able to complete their plans and have them certified.

Cities and regions involved in bioterrorism preparedness, therefore, are beginning to think through medical management issues as they develop their local response plans. Most locales have included procedures for acquiring additional resources and equipment, either through mutual aid agreements or from *federal stockpiles*. However, very few have considered planning for additional medical staff. Aside from resource acquisition, the primary challenge will be treating patients in a timely manner while maintaining a high level of care. Decisions will have to be made about who can be discharged, when they will be discharged, and what types of long-term care they will receive.

Hospitals also remain unable to cope with even a limited *bioterrorism event*. Few hospitals have more than 100 beds available at any given time. Similarly, hospitals keep no more than a few weeks of medication on hand, which would be inadequate to meet the needs of a large number of casualties.

Most locales have triage plans and have identified schools, churches, arenas, conference spaces, and other sites where large numbers of people can be accommodated and processed or prophylaxed. Many locales have also considered setting up triage functions outside of hospitals, but most hospitals are reluctant to expend planning resources to coordinate triage activities given other health care priorities. Staffing and supplying triage points will be a significant challenge, especially as the numbers of exposed persons rises. Aside from staffing and supplying triage points, the primary challenge of triaging large numbers of people will be processing them in a timely manner.

In addition to the NDMS and the MMRS Contracts, OEP began stockpiling vaccines and drugs for use against biological agents in 1998. The stockpile program, managed by the Department of Veterans' Affairs, initially targeted smallpox, anthrax, tularemia, and plague, with other diseases to come later. While this program allowed for a greater potential availability of needed vaccines and antibiotics, a recent General Accounting Office (GAO) report indicates that the efficacy of the these stockpiles is compromised by poor management controls and a lack of required items. A physical inventory of OEP's stockpiles indicated a discrepancy of more than 12 percent, and also revealed an alarming number of expired items that sometimes included the entire supply of a specific drug. According to the GAO, the principal cause of these problems was a failure to implement basic internal controls that could reasonably assure that all medical supplies and pharmaceuticals are current, accounted for, and available for use. In fact, the inventory systems lacked basic information required for good record keeping, such as documentation and back orders, replacements, and shipment and receipt of all pharmaceutical and medical supplies. Furthermore, the OEP did not conduct periodic inventories, maintain program policies and procedures, or provide adequate security of the *stockpiles*. As a result, GAO concluded that the system cannot be relied on to provide complete medical support in the case of a bioterrorism incident.

In response to the perceived need for a more efficient, centralized supply of drugs and vaccines, the Centers for Disease Control received $51 million in FY 1999 to establish the National Pharmaceutical Stockpile (NPS) Program. The "*stockpile*" will consist of two distinct elements:

Pre-positioned Stockpiles of Medicines, Equipment, and Medical Supplies

Full packages will be kept at Louisville and Memphis, the major hubs for Federal Express and other parcel carriers. One package each will be also kept in Dallas, Sacramento, and Philadelphia, while a half package will be kept in Hawaii and another in Puerto Rico. Currently, the stockpile is focused on compiling antibiotics, mainly oral and intravenous ciprofloxacin and doxycycline, and medical supplies like ventilators, tubing, needles, and so on. As of today, the push packages. do not contain vaccines. Each push packages is identical in size and contents and designed to provide a rapid response capability. The combined push packages will contain sufficient quantities of antibiotics to provide prophylaxis to 7.3 million people for three days. Each package is located and maintained to reach any part of the country within 12 hours after their release by the Director of the CDC. Under the program, CDC will assemble the push packages with the Department of Veterans' Affairs assigned with management responsibilities. To address concerns regarding tracking and inventory controls, CDC and the VA are installing a computerized inventory management system that will track contents, shelf life, and movement of materials into and out of the packages.

Vendor Managed Inventories (VMI)

VMI are designed to support the maintenance of excess capacity by the pharmaceutical industry. As opposed to the prepositioned supplies provided by the eight NPS push packages, the VMI are designed to provide follow-on supplies from the nation's pharmaceutical supply system as push package supplies are exhausted. Through contracts with HHS, certain drug manufactures will agree to maintain excess production and supplies and make them available to the government in an emergency. This reserve provides a source of additional antibiotics and materials necessary for long-term treatment and prophylaxis.

While the NPS ensures that certain drugs and vaccines are on hand in the case of a biological or chemical emergency, it does not ensure that these drugs will be available to local and state hospitals when they are needed. In fact, most officials estimate that drugs received from a regional stockpile will not be available at the local points of distribution for 36 to 48 hours after an incident has been detected. As a result, local health authorities recognize that they must prepare for an unexpected situation that requires large amounts of *antibiotics* or *rare vaccines*. Those physicians and emergency responders involved in *bioterrorism preparation* for the 1996 Atlanta Olympic Games stated that placement of necessary pretreatment drugs in local medical facilities was an important lesson. Despite the availability of CDC's regional antibiotic stockpile, hospitals and local public health departments need to identify and augment supplies. During the May, 2000, TOPOFF exercises in Denver, participants underlined major problems with prophylaxis. During their notional deployment in the

Topoff scenario, the push packages were delivered nearly 48 hours after they were requested. Moreover, no system was in place to deliver the antibiotics from the airport to the city, and no clear method of distribution to the populace was available. Similarly, because the distribution of antibiotics was delayed, few serious cases of disease were actually prevented. While all of these TOPOFF findings are notional given the artificial nature of some of the exercise play, these findings nevertheless highlight key issues that need attention.

Although the options discussed above could potentially be effective for distributing medical supplies during a bioterrorism crisis, the details of the necessary public-private partnerships have yet to be worked out. Even in the public sphere, drivers and the most optimal supply routes for distributing massive amounts of medication have not been fully identified. Nor have local DoD assets and National Guard assets (other than the Civil Support Teams which have little distribution ability) been integrated into planning for distribution purposes.

Although the CDC has put a lot of effort into ensuring that logistically the NPS push packages can be loaded on C-130s air transports and shipped anywhere in the contiguous United States within 12 hours (ideally the push packages could be delivered in under 8 hours—forty minutes for loading time, one hour and thirty minutes for flight preparations, and 4-6 hours maximum flight time), once the push packages are delivered to a local airport, control is taken by local authorities. Federal planning fails to plan for the integration of federal and local assets by operating under the assumption that once the push packages are delivered, local distribution mechanisms have the capacity to distribute them.

In reality, the medical distribution plans of most locales are rather vague and do not detail how current medical supply mechanisms or supplemental mechanisms would actually operate in the event of a *bioterrorist attack*. Rather, they simply identify potential partners in achieving this task. Likewise, while Emergency Medical Disaster Plans note that emergency medical service personnel and health care facilities should provide mutual assistance in situations in which local resources are overwhelmed, few formal partnerships have actually been put into place. Procedurally, these plans form the basis of cooperative agreements that could be applied during a bioterrorism event, but operationally they lack mechanisms for identifying where resources are housed or ways of transporting resources between facilities. While cities are developing plans for prophylaxis distribution systems, overcoming the local manpower shortages produced by the response to a bioterrorism incident may require state and federal assistance. OEP has done a good job of organizing specialized national medical response teams that could be deployed to support the medical response to a bioterrorism incident. These teams, however, are oriented toward providing additional treatment capacity and are relatively small in size; thus, their role in supporting mass prophylaxis programs is likely to be minor.

As was mentioned previously, the Department of Defense has also focused on developing and maintaining relatively small WMD response teams like the Marine Corps' Chemical and Biological Incident Response Force (CBIRF) and the National Guard's WMD Civil Support Teams. CBIRF consists of approximately 300 personnel skilled in the provision of security isolation, agent identification, and medical support. As with the MMSTs, CBIRF teams may represent a useful resource in the event of a significant number of casualties. However, most WMD Civil Support Teams are trained and equipped for agent recognition and environmental clean up, not for providing medical care. As such, their contribution to patient treatment is questionable. Given their small complement, these specialized teams are not likely to provide significant assistance in supporting prophylaxis. Neither OEP nor the various civil support planning elements at the Department of Defense have focused adequate attention on planning how their available manpower resources might be used to support large-scale prophylaxis programs. DOD, both within regular and National Guard units, has the potential for providing large numbers of personnel to support prophylaxis programs.

Key Issues and Recommendations

Triage

A pre-incident plan for the location of triage stations needs to be in place before an event. This plan should include multiple points to reduce congestion, by both vehicles and people, in a single area and keep management of the area to a realistic size. Arrangements should also be made for triaging patients outside of hospital facilities and identifying both medical and non-medical triage personnel.

Providing public information will be an important tool for encouraging "*self-selection triage*" as well as for directing the public to appropriate triage points. A public information strategy should be devised in advance of an attack that will provide clear and accurate information about when treatment is required and where it should be received. An effective public information campaign can inform people whether treatment is necessary or not given their particular situation and direct them to the appropriate treatment center. Such a campaign can greatly reduce the stress that hospitals and other treatment facilities and triage points will feel.

Arrangements for counseling patients should be made to reduce panic and confusion at triage points and to ensure order. A follow-on mechanism should also be put into place to monitor patients' mental and physical conditions and perform secondary triage accordingly. It is unlikely that patients at triage or other prophylaxis and treatment points will accept whatever treatment is offered to them without asking any questions. Rather, patients will be concerned given the novel and frightening nature of bioterrorism. Patient-counseling services can calm and reassure victims, help maintain order, and reduce confusion over how and why treatment regimens are being administered, especially self-treatment regimens.

Legal constraints forbid non-medical personnel to diagnose, treat, or refer medical patients. But in the event of a bioterrorist attack, allowing non-medical personnel to perform some of these functions on a limited basis may be essential if there are large numbers of people presenting for treatment. At present, no agreement has been reached on which non-medical personnel could make what medical decisions should it become necessary. Agreement must be reached in advance about who can make triage decisions in the event of a bioterrorist attack, and the medical and legal ramifications of these decisions need to be considered. Giving the authorities the option to suspend legal action for non-medical referrals during times of crisis may give triage personnel more security in the decisions they make.

Pre-Event vs. Post-Event Prophylaxis

Given the manpower requirements of prophylaxis systems, emergency medical services, medical care, security, and public safety, some thought might be given to providing certain segments of the population with pre-event prophylaxis. This might be most appropriate with anthrax vaccine. Pre-event prophylaxis will ensure the availability of key segments of the population—fire and police services, emergency medical services, and medical care professionals—by protecting them from certain agents, thus eliminating any psychological impact that may prevent their involvement with response activities. A number of issues need to be addressed before undertaking such pre-event prophylaxis programs. The first issue is the cost of such a program. The Department of Defense has provided over 1.7 million doses of anthrax vaccine to over 440,000 U.S. service members at a combined cost of over $150 million.

According to some estimates there are approximately 2.8 million fire, police, and EMS personnel who could be involved in bioterrorism response. This figure does not include public health personnel, pharmacists, laboratory technicians, and emergency room doctors and nurses. Providing anthrax vaccinations to fire, EMS, and emergency room doctors and nurses in just the 50 largest U.S. cities would be exorbitant. The second, and related, issue is the safety of these vaccines. As the Department of Defense's experience with anthrax vaccine demonstrates, there is a need for reliable production entities with stringent quality control systems in place.

In addition, there is the need for research to ensure no adverse side affects are associated with the specific vaccine. Meeting these two requirements, while providing education about the vaccine to those receiving it, should reduce the number of people refusing to participate in the program due to safety concerns. This is a key consideration when participation in this inoculation program will necessarily be of a more voluntary nature than the military's *anthrax vaccination* program. The final issue is the likely response from the general public when a pre-event prophylaxis program begins.

Some members of the public are sure to feel anxiety regarding their continued vulnerability to an attack and could begin calling for a program to provide prophylaxis to the general public, substantially increasing the cost of such a program. Calls to increase the availability of vaccines might come from those individuals who have already been prophylaxed, but who fear their family members remain vulnerable. With justification, they may call for vaccines to be provided to their families or else they may remain with their family members after an attack.

Scalable Prophylaxis and Treatment Capability

A *catastrophic attack* producing thousands of casualties is not the highest likelihood contingency. This suggests localities should develop prophylaxis and treatment plans that are phased or broken into escalatory segments. For situations in which small amounts of agent is released, moving from the normal situation to executing a plan to provide millions of doses or prophylaxis or treat 5000 casualties is not appropriate. The possibility of high-end attacks must not be ignored, but they are not the only contingencies against which planning should occur. A "*scalable*" plan is required to rapidly begin providing treatment and prophylaxis and then scaling the system upward if required. This flexible approach provides the capability to treat and prophylax in an effective and efficient method.

Such a system may require a combination of distribution methods. For example, a limited canvassing system to deal with lower impact, higher likelihood events, a local PODs capability for dealing with large events, and finally a "*full-up capability*", including state and national resources in the event of a catastrophic event. It may also entail a scalable, phased approach to providing medical treatment—with hospitals treating the first 100 to 200 casualties, locally established and maintained alternative care facilities treating up to 500 casualties, and then state and federal assets providing medical care to casualties above 500. Building flexibility and scalability into response plans makes the process more difficult with plans becoming more complicated, but it will serve to maximize resources.

Reliance on Federal Stockpiles

Another important issue is the almost total reliance on federal stockpiles to provide medical supplies. Such reliance could diminish the timeliness of the response given the rough estimations for the time between the release of national stockpiles and the arrival of those supplies at local points of distribution. Push packages are slated to arrive locally within 12 hours after they are released. Once they have arrived, additional time will be needed to break down the package, a task that, according to one jurisdiction's bioterrorism plans will require an additional 12 hours if 32 person crews are utilized in 2-3 hour rotations. Once broken down, more time is required to move supplies from their arrival point to points of distribution, determined by the distance between the arrival point and the distribution points, availability of transportation, and any obstacles like traffic.

Local assets will have to provide the necessary supplies until this external assistance arrives. But few hospitals or primary care clinics possess a sufficient amount of drugs and supportive medical equipment on-hand for treating or providing prophylaxis to an unusually high number of patients. In the event of a biological attack, health facilities could be overwhelmed, lacking bed space, medicine, protective gear, and even staff. According to an administrator at a major urban hospital, "as it is, emergency rooms are limited in what they can do for many casualties, especially if it is a mass casualty situation. If we had 100 casualties, it would severely stress the hospital system." Addressing this problem is a difficult challenge. One option that has been suggested is that localities should be provided with the resources to develop an independent first-tier prophylaxis capacity with local medical care providers, either through mutual aid agreements or regional buy-ins, to allow for a comprehensive response in the 48 hours before federal aid arrives.

One way for smaller communities to combat the logistical difficulties of acquiring and stockpiling both vaccines and antibiotics is to create, in essence, a "*supply bubble*" within the current hospital pharmacy system. Implemented in New York City, this method involves purchasing more drugs than the usual one-week on-hand supply, but fewer drugs than would be used by the end of their *shelf life*. Such an approach, however, is questioned, both for its feasibility and cost-effectivness, particularly for smaller metropolitan areas that are not likely to have the necessary financial resources. Another approach may be to build on the OEP stockpiles, assuming that the problems associated with that program identified by the GAO are aggressively resolved. To ensure efficient use of local resources, a system should be established to track the movement and use of medical supplies during a bioterrorism response. A tracking system will give public health officials some awareness of how medical resources are being used so they can plan accordingly for procuring the necessary supplies as they are depleted. A tracking system will also ensure that patients are directed to facilities where supplies are still available. Without some type of tracking system in place, officials will quickly lose control over the flow of medical supplies.

Contents of the National Pharmaceutical Stockpile

Clearly, it will be necessary for the federal government to provide most of the antibiotics, vaccines, and associated supplies necessary for prophylaxis and treatment in response to a *bioterrorism incident*. This is the reason for the NPS Program. Keeping in mind the likely size of groups requiring treatment or prophylaxis, the variety of agents for which the stockpile should be prepared, and the corresponding variety of materials, what types of and how much medicine should the stockpiles contain? It is impossible to develop a stockpile large enough to prophylax the entire population of the United States against the entire range of threat agents. The size and the contents must balance the competing interests of cost, efficacy, and comparative likelihood of certain events, while recognizing the degree of risk associated with choosing not to produce a certain medicine or vaccine or produce it in limited quantities. As mentioned previously, the combined *push packages* will contain sufficient quantities of antibiotics to provide post-exposure prophylaxis to approximately 7.3 million people for three days. The types of antibiotics being stockpiled are appropriate given the high degree of commonality between treatment modalities for agents. The amount being stockpiled, however, might not be sufficient.

Prophylaxing the population of a city the size of New York would exhaust the antibiotics supplies of all eight push packages within three days. This does not account for antibiotics necessary for treatment and does not provide additional antibiotics for attacks in other parts of the country. Of course, New York is one of the largest U.S. cities, and most terrorists are not likely to disseminate sufficient quantities of anthrax, plague, or tularemia to infect the entire population of that size. But without good epidemiological data indicating the scope of an attack, there is a risk that a substantial portion of a city may be exposed; prudence will dictate applying prophylaxis to more people rather than fewer. The size of the antibiotics portion of the push packages appears to be appropriate given the low probability of a catastrophic attack, but only if adequate detection and assessment capabilities are in place. Early detection and accurate assessments will ensure the size of the prophylaxis effort is calibrated according to the size of the attack. While the current amount of antibiotics provides a hedge against a catastrophic event, the antibiotics portion of the packages might be reduced as confidence in the nation's detection and assessment capabilities increases.

The other issue is the lack of vaccines in the *stockpile*. Other than possibly reducing the course of antibiotic treatment necessary to prophylax successfully, anthrax vaccine does not provide effective post exposure prophylaxis. This fact, in combination with its lack of person-toperson transmissibility and its 20-30% mortality rate even after treatment, makes stockpiling anthrax vaccine unnecessary. Because it is contagious, some quantities of smallpox vaccine will be necessary as a hedge against possible future outbreaks, relatively unlikely as they may be. The issue is how much. Re-immunizing the entire country against *smallpox* would prevent all future infections, but the cost of such a program would be prohibitive, especially when the chance of a *smallpox attack* is minimal. Stopping short of providing smallpox vaccine to everyone in the country, the current strategy is to stockpile sufficient quantities of vaccine to immunize the entire population surrounding the affected area to prevent new infection and prevent its spread outside that immediate area. Roughly 9 million doses of vaccine are currently available, but only by diluting current stocks. In FY 2000, CDC is spending over $16 million on the special, bifurcated needles required for immunization, dilutant, and additional immune globulin. Because the effectiveness of the current vaccine is in doubt, CDC is spending an addition $22.5 million for research and production of a new vaccine.

There is considerable risk associated with this approach. The effectiveness of providing vaccine to only a limited segment of the population depends on detecting the smallpox outbreak before it spreads outside a confined geographic region, like a metropolitan area. Providing vaccine to the population of that area can prevent further transmission between members of that area's population. As the disease spreads within an increasingly larger area, the number of people needing vaccine increases. Given the mobility associated with urban life-styles and current gaps in the nation's surveillance capabilities, a smallpox outbreak could spread outside a limited geographic area before it is detected. Such a situation makes a major smallpox outbreak a complicated event in which vaccine delivery will become only one aspect of the fight to stem it. Looking at such a scenario also highlights how difficult it is to determine just how much smallpox vaccine should be on hand across the United States.

As a hedge against a possible smallpox incident, the current supplies of vaccine should be readied for a rapid, mass prophylaxis program and a surge vaccine production capability should be established. One or more vaccine producers would be compensated for maintaining a reserve surge capacity for future response and for the cost of production and any associated disruption if a smallpox incident occurs. This surge capability should be able to manufacture additional vaccine if current reserves are insufficient for response to an incident. At the same time, efforts should be made to develop measures of merit to help determine the levels at which the national smallpox stockpile should be maintained. Those measures should make it clear to policy makers and others why a particular level of stocks is optimal.

Finding Manpower

Staffing is a challenge that for the most part must be met at the local level, yet to date few localities have incorporated staffing issues into their planning. As previously discussed, several options exist for acquiring additional staff, including establishing mutual aid agreements with local hospitals, developing strike teams made up of local or regional medical personnel, or calling in national response teams. The efficacy of each of these options has already been detailed, but it is clear that a great deal of coordination will be necessary to integrate any one option into the local response framework. As such, additional staffing must be considered when creating response plans if public health and medical personnel are to effectively treat patients and provide prophylaxis. More planning is needed to determine how the state and federal government can provide additional manpower in support of prophylaxis and treatment as a hedge against large scale bioterrorism incidents.

As was previously discussed, canvass-type distribution systems require massive numbers of people. Manning specially designed points of distribution also requires large numbers of individuals. This challenge is less daunting for those plans utilizing existing pharmaceutical distribution systems. Pharmacists and pharmacy technicians already in place at existing pharmacies, drug stores, and clinics would form the core of the prophylaxis workforce. But even they are going to require additional resources to handle the much higher than normal demand at their stores. They will also need security personnel to protect their persons and supplies. Given the sheer numbers that may be involved, finding people is a challenge to meet with local resources. New York has developed two models for rapidly prophylaxing a large exposed population in the event of an attack. The first model, based on the canvass approach, calls for bringing medications directly to the home; with an exposed nighttime population of 7.5 million people, delivering a five-day supply of *ciprofloxacin* would require 41,652 personnel.

The second model, based on the points of distribution approach, would require 45,344 personnel to dispense 75 million tablets in 48 hours. This type of distribution system might be required in situations in which people should remain quarantined, the disease is contagious, or the rate of infection is so high that there is no alternative to self-treatment. However, while New York's stockpile program could serve as a model for local entities, it is extremely expensive and labor-intensive, and would require an enormous level of logistical support. Furthermore, as TOPOFF demonstrated, personnel charged with delivering medication cannot work as quickly as the prophylaxis models indicate, resulting in a longer, slower distribution process.

Improvements in Prophylaxis Plans

Identifying where POD sites will be located, the best transportation routes for distributing medical supplies, and available personnel should be done before an incident to ensure that the distribution system is operational. Improved evaluation methods must be developed to assess the adequacy of local prophylaxis plans developed through the bioterrorism track. Most of the plans have not been operationally tested due to the cost and disruption that would be imposed by a large-scale exercise. Smaller exercises could be done, but their ability to evaluate the entire plan is limited. One tool that might help to evaluate these plans is a computer simulation to model how a locality's prophylaxis plan would be implemented and how well it would work. New York City's models could be adapted by other cities to more accurately determine their own personnel and material requirements before a formal plan is drafted.

Such computer simulations might be provided by some of the bioterrorism response system architectures being developed at the national laboratories and the Defense Threat Reduction Agency. A key challenge is the time needed to move personnel to the affected area. Local plans for prophylaxis should recognize and account for additional state and federal assets that will be made available by planning for how these assets will be utilized.

Likewise, state and federal assistance plans should account for the local prophylaxis plan by accounting for the type of system in place, the work already accomplished by the time of their arrival, and how they can be integrated into the local system. OEP, although it does not possess an organic manpower resource pool, has the potential for organizing systems by which additional state and national medical manpower resources could be exploited to support prophylaxis response programs. This could include asking local and state officials to develop mutual aid agreements to provide additional personnel to support prophylaxis. It should also include systems for national mutual aid protocols. While the CDC has worked out the logistics behind delivering the NPS to localities, once there, the CDC turns over responsibility for distributing the NPS to local care facilities. Few arrangements have been made locally for doing this.

5

Disease Control and Prevention

The Bioterrorism Preparedness and Response Program (BPRP) office at CDC has a unique vantage point from which to improve the nation's bioterrorism preparedness. It serves as the main interface between federal, state, and local entities on issues of public health and medical preparedness for meeting the bioterrorism challenge. It is charged not only with coordinating the bioterrorism preparedness activities of various entities at the CDC, but also with leveraging federal resources to improve bioterrorism readiness in state and local public health systems. To these ends, dramatic improvements have been made over the past two years. The BPRP has begun the process of raising awareness about the role that the public health and medical communities play in preparing and responding to bioterrorism and has helped integrate them into the emergency response systems at the federal and local levels.

The following section delineates issues pertaining to BPRP and the CDC's current strategy for building bioterrorism preparedness, their internal organization and structure, and how each of these has affected the current state of preparedness. It also provides recommendations and potential next steps.

Planning Assumptions

CDC's planning assumptions have been based on single-factor, worst-case, lower-probability scenarios. CDC defines the bioterrorism threat by identifying the agents that have the greatest potential for producing the highest number of casualties and fatalities, not on the comparative likelihood of such a scenario occurring. CDC measures an agent's potential for producing high numbers of casualties and fatalities through a combination of ease of dissemination and transmissibility, high mortality, potential for panic, and impact on the public health and medical system. CDC's initial threat assessment, while recognizing a fairly broad range of bioagents, gave priority to a relatively small set of threats without accounting for other factors in their assessments such as difficulties associated with acquisition, production requirements needed for producing mass casualties, or dissemination considerations.

By focusing on a narrow range of outcomes, CDC assumptions do not account for the relationships between the components involved in achieving a mass casualty attack or the difficulties that terrorists may face, and the capabilities they will need to bring these components together effectively. They also do not account for the range of terrorist motivations or the relationship between motivations and mass casualty scenarios. This report emphasizes the need to examine the threat as a complex set of factors, including the relationships between terrorists' motivations and the number of casualties desired, and the technically feasible approaches needed to produce mass casualties. Recognizing that it is vastly more difficult to mount an actual response to worst-case scenarios, focusing on only a relatively small set of contingencies—admittedly lower-probability, higher consequence—simplifies bio-terrorism planning and preparedness requirements by narrowing the range of contingencies. But it can also drive suboptimal and possibly very costly response options.

Preparing the nation's public health and medical system to detect and respond to the full range of more likely medium and low-end scenarios is more difficult. CDC seems to have assumed that preparing the nation's medical and public health system to respond to worst-case scenarios will provide a capability to respond to the full-range of scenarios. This assumption is valid across all functional areas. For example, expending large amounts of financial resources on research, development, production, and stockpiling of *smallpox* vaccine improves response capacity for a single agent capable of yielding massive casualties, but it does not promote further enhancement of tools for improving the response capacity for a wide array of agents that might be useful in contingencies of lesser consequence but greater probability. What is the proper balance in this case between the need to be adequately prepared for more likely contingencies and the need to hedge against high-consequence ones? Another example is providing reagents to laboratories for a narrow set of agents that does little to ensure preparedness for the full-range of potential agents that may be used. Finally, the emphasis that has been placed on agents with the greatest potential for producing mass casualties has resulted in distorted perceptions within the response community, policy community, and general public of what a bioterrorist attack is most likely to look like. Such a perception remains problematic, partly because the response communities tend to see the response to bioterrorism as "*all or nothing*", partly because policy makers have an inaccurate conception of the best use of resources, and partly because the general public may overreact to a bioterrorist attack that in reality was localized and ineffective.

The National Pharmaceutical Stockpile, a central activity in building public health and medical preparedness for bioterrorism, provides an example of the kinds of questions that can be raised when planning is based on "*worst case scenarios.*" While everyone agrees that a bioterrorist attack with smallpox is a potentially devastating event, a number of factors mitigate the probability of such an attack being conducted. Issues of terrorist access to smallpox, for example, or the danger to a terrorist in trying to use it are factors that diminish the likelihood of a *smallpox attack*. This raises serious planning questions regarding initiatives underway for producing and storing *smallpox vaccine*. Is the production and stockpiling of vast amounts of smallpox vaccine a wise use of resources given the low probability of such an attack? Although some hedging against a smallpox attack is necessary—for example, preparing current vaccine stocks for deployment, purchasing needles and globulin—are research and development efforts for smallpox vaccines currently underway a cost-effective priority of preparedness building?

Some people would argue that it is, contending that some vaccine is necessary, and, if a little is made, it is not significantly more expensive to make considerably more. Particularly since the smallpox vaccine, as one expert put it, is "essentially out of the tech base already," money spent in this area as a hedge against the kind of truly catastrophic event that a smallpox outbreak would represent is not an unwise resource allocation.

The question, of course, is not whether or not smallpox vaccine should be produced. Rather, the issue is how much and at what opportunity costs. HHS recently entered into a 10- year contractual agreement to spend approximately $40 million dollars per year to research and produce additional quantities of smallpox vaccine. Money spent on smallpox vaccine cannot be spent on other areas of *bioterrorism preparedness*. Without robust detection and assessment capabilities, for example, early quarantine and treatment measures that might contain an outbreak are not likely to occur. Moreover, surveillance, epidemiology, and laboratory capacity are crucial to responding to all varieties of *bioterrorism scenarios*. A stockpile of smallpox vaccine is useful for only one type of scenario. Unofficial figures place the costs of preparing existing doses of smallpox at $17 million, probably a good hedge. But the $27 million that is already committed for research and development into better production techniques for smallpox vaccines might be better spent stockpiling pharmaceuticals able to treat victims for a wider array of bioterrorism attacks or to build epidemiological and microbial forensics capabilities to assess the nature of an attack and aid law enforcement.

If such an effort could lead to cheaper and more cost effective vaccine production, however, the investment might be worth it, but the cost savings would have to be substantial. Similar questions arise about research and development for activities for *anthrax vaccines*. Post-exposure vaccination for *anthrax* is not an effective treatment, and pre-incident vaccination of critical response personnel is prohibitively expensive. Given the complexity involved with using *anthrax*, *botulinum toxin*, *plague*, and *tularemia* effectively as weapons for inflicting mass casualties, what is the appropriate size of a pharmaceutical stockpile for prophylaxis and treatment of exposure to these agents? The push packages currently provide the capacity to prophylax just under eight million people. They are largely comprised of *cyprofloxicin* and *doxycycline*, two antibiotics with the widest applicability to *bioterrorist agents*. This capability, combined with vendor managed inventory and state and local supplies or caches, is sufficient to prophylax and treat victims of the higher probability bioterrorism attacks that are localized and limited exposure attacks.

Some additional consideration should be given to identifying antidotes for other bioterrorism agents and stockpiling some level of reserve stocks to be shipped to an attacked area, even if these reserves are as simple as anti-diarrheal medications for countering the effects of *salmonella* or *E. coli*. While medications are an important part of preparedness, the key to using these medical resources effectively will be the quality of the tools that are available for assessing the scope of an attack. Good assessment tools can reduce the amount of medication that needs to be stockpiled by ensuring that those resources are properly directed and used efficiently. Therefore, given current pharmaceutical supplies, the best use of resources appears to be: (i) capacities that are applicable to a wide variety of bioterrorist attacks; (ii) assessment tools that can appropriately guide the response; and (iii) capacities that have spin-off for responding to naturally occurring infectious disease outbreaks.

It is unclear whether the CDC has sufficiently outlined the bioterrorism detection/assessment and response requirements the public health and medical system should be able to meet. The recently released recommendations of the CDC's Strategic Planning Workgroup are a good first step toward a better definition of the core capacities for pubic health and medical bioterrorism response capabilities. The grant programs and the strategic plan do a good job of identifying the functional areas within the practice of public health that are necessary components of bioterrorism detection, assessment, and response. They also identify the types of activities that must be taken to develop these functional capabilities. What they fail to establish adequately are well-defined operational requirements that together constitute measures of merit for an effective response. Such measures of merit are needed for each of the key functional areas. Such measures are important because they provide a metric for evaluating the progress of preparedness programs. Also needed is a measure for the success of integrating these components into a system. Once these measures have been established, current programs should be linked to specific operational requirements and time frames for meeting them. Without clear requirements and related measures to evaluate progress, assessing preparedness programs over time and making adjustments when necessary is difficult.

The BPRP Office should be given the necessary leverage to manage the direction of, and integrate, public health and medical bioterrorism programs at the CDC. When the BPRP was established nearly two years ago, it was placed in the National Center for Infectious Diseases (NCID), a natural choice in that bioterrorism is a problem of infectious diseases, but also an odd choice given that BPRP's mandate was to coordinate bioterrorism planning and preparedness activities across CDC's various centers. This choice has made the BPRP essentially one program of many within the NCID, making it difficult to coordinate the activities of programs with similar status in other centers. For example, the National Center for Environmental Health's national pharmaceutical stockpile program, the Public Health Practices Program Office's Health Alert Network (HAN), and the Epidemiology Program Office's detection and surveillance activities are first and foremost responsible to the directors of the respective centers or program offices that house them and not to BPRP, which is tasked with coordinating them. Given BPRP's status within the NCID, it is difficult for it to exert leverage over other program offices in other centers.

This challenge is exacerbated by the fact that individual centers directly receive funds for their bioterrorism-related programs. While NCID receives some congressional funding to support BPRP, much of its budget is generated out of the bioterrorism budgets of the other centers (i.e. the HAN or National Stockpile). This essentially makes BPRP beholden to the programs they are expected to coordinate. In order for BPRP to manage public health and medical bioterrorism preparedness activities more effectively, it must be given more leverage as well as some control over budgetary decisions. One way to do this is to elevate the BPRP to the level of the Office of the Director at CDC and give BPRP budgetary authority over the BT preparedness programs at CDC.

Through BPRP's grant program, CDC has taken an even, uniform approach to improve response capacity within the public health system. To a great extent, the CDC's approach to building capacity is the result of a sense of urgency that a bioterrorist event that produces catastrophic levels of casualties is "not an if, but when" scenario. In order to address this imminent and overwhelming threat, funding was freed through emergency appropriations to improve functional areas that are weakest in the public health departments across the country. The "fix the weak link" approach was adopted over an integrated system approach because many felt that the "*bioterrorism buzz*" and corresponding flow of financial support for preparedness programs could wane in subsequent years. Therefore, attempting to develop a systems approach to the problem and implement a multi-year, phased preparedness strategy was futile. Given these concerns, limited resources, and time constraints, distributing available resources to as many states as possible to support as many functional areas as possible made sense.

In allocating its grants, BPRP has simultaneously improved all five major components of the public health systems' response capability. The grants program has divided the funding fairly evenly between detection/assessment tools—surveillance, epidemiology, laboratory—or the "*front-end*" of the system, and response tools—establishing the pharmaceutical stockpile or research and development into vaccines—or the "*back-end*" of the system. Furthermore, substantial funding has been allocated to other agencies within HHS for building response capacities—the MMRS and Bioterrorism Response Plans. But this approach suggests that both detection/assessment tools and response tools should be of the same importance at this stage of preparedness efforts and that each makes a comparable contribution to an effective bioterrorism response in the current circumstances.

Is that assumption warranted? Probably not, given that many of the front-end tools are essential for detecting an attack, characterizing its nature, guiding the response, and ensuring that response resources are efficiently utilized. Therefore, to a large degree the effectiveness of response tools will largely depend on the quality of the detection and assessment tools. For this reason, the "*front-end*" of the system should be given priority consideration given that it can enable the response system.

Planning Coordination

Senior HHS leadership must better coordinate CDC's Bioterrorism Preparedness and Response Initiative with the Office of Emergency Preparedness's disaster preparedness initiatives. Both BPRP and OEP provide local agencies and departments with federal funding through grant programs as well as guidance on grant implementation, but each organization focuses on different aspects of bioterrorism preparedness. CDC concentrates on strengthening the public health system at the state and local levels to improve *detection* and *assessment capabilities*. OEP's all-hazards approach, which includes chemical or biological terrorism incidents, is oriented toward improving the *medical management system* of localities. As these programs have progressed, it appears that certain aspects of the CDC and OEP programs have begun to overlap and are proceeding toward the same goal without the benefit of coordination between them.

Cleanly dividing responsibility for the public health system from the medical system is problematic in both theory and practice. While the nation's medical system plays an important role in providing treatment, it is also key in both surveillance and initial assessments. The CDC Strategic Plan for Biological and Chemical Terrorism Preparedness and Response includes medical treatment and prophylaxis in a comprehensive public health response to biological or chemical terrorism. CDC is also compiling the National Pharmaceutical Stockpile and is planning to provide on-site consultations for both medical treatment and prophylaxis. At the same time, OEP has asked local terrorism planners to develop bioterrorism preparedness and response plans through the MMRS grant process. Although these bioterrorism plans are focused on development of local response systems,

OEP is also asking planners to include surveillance, epidemiology, and laboratory response components in their plans.

Although some overlap of efforts is inevitable, senior HHS management must ensure that the delineation of bioterrorism preparedness roles and responsibilities for BPRP and OEP is done in such a way that it maximizes the contribution of both entities. Solving this issue could require organizational changes, including changes in budgetary and management responsibilities at HHS, as well as shifts in program responsibilities and structure. One option could be to specify that one office—BPRP, for example—has the overall leadership role and that OEP's role is clearly one of support. Changing the current status of responsibilities will be difficult, but it may be the best way to facilitate the programmatic decision making that will be key to the success of the effort over the long term.

Development of a bioterrorism preparedness "*urban strategy*" gives HHS leadership an opportunity to address these concerns while providing local planners with a set of guidelines and suggestions that encourage and facilitate local planning and capacity building. CDC should develop and disseminate such an "*urban strategy*" bioterrorism planning and organization template designed for use by local public health departments and other local entities. Absent such a template, local planners lack a reference for defining overall preparedness tasks and activities, identifying system requirements, listing potential partners and measuring progress, while CDC lacks an important tool for ensuring that local agencies effectively leverage bioterrorism support into improved capabilities. Because effectively designing, organizing, and implementing a local detection, assessment, and response system involves a wide array of other organizations and entities, both public and private, the template should include sections on:

1. Organizing local health surveillance systems, including the integration of nontraditional partners;
2. Developing bioterrorism response plans based on the previously discussed tiered approach;
3. Ensuring effective and integrated emergency communication capabilities;
4. Developing and managing local training programs;
5. Coordinating with the public safety and law enforcement communities;
6. Developing linkages with other localities and the state public health department.

In addition to developing an "*urban strategy*" and related template, CDC should provide local public health departments with financial support to design a local bioterrorism system based on the template, develop a plan for creating their system, and then execute it. BPRP's plan to create a bioterrorism coordinator for each of the 50 states represents a good first step. Using the state coordinators to link BPRP with local health departments, planning grant recipients should be required to specify objectives related to each of the items listed above and develop operations concepts to achieve them. While planning and organization provide the foundation, CDC must also allocate sufficient resources to build local capacity for detection, assessment, and response. Grant recipients should begin with city and county public health agencies but also include hospitals, clinics, and other local partners.

While CDC is working closely with national public health organizations, many members of the public health community at the state and local level have expressed apprehension regarding CDC's proclivity to provide operational guidelines and information without being sufficiently receptive to suggestions or concerns from state or local public health officials. One example that has been mentioned frequently is CDC's reluctance to allow laboratory technicians to alter laboratory testing protocol text because of the requirement for a standard protocol to facilitate criminal prosecution. In addition, some states have asked for greater flexibility in grant execution, specifically the ability to move funds between different focus areas which would facilitate grant execution in the context of state or local procedures for hiring new employees or obligating funds.

Many localities feel that their suggestions have received a lack of attention and perceive a top down approach to preparedness building. But given the large number of entities with whom CDC must deal, it is easier for an organization of CDC's size to disseminate information to a large number of people and organizations than it is to receive and assimilate information from the same number of recipients. This tendency is only reinforced by a lack of personnel within the Bioterrorism Preparedness and Response Program Office, given the scope and importance of the work it does.

In administering the federal grants process, BPRP has played an integral role in coordinating and providing guidance to local public health entities and coordinating national organizations and associations with expertise on the BT challenge, but more could be done. Much of the time BPRP's small staff is consumed with providing guidance to local entities on the grants process, making BPRP largely an administrative office for federal grants rather than a central focal point of all bioterrorism-related public health and medical activities. Instead, BPRP relies heavily on national organizations to provide guidance on more technical issues related to preparedness, a reliance that results in the dilution of BPRP's overall coordinating function and often sends mixed signals to local entities about where their first point of reference should be.

BPRP has done an excellent job in coordinating with national organizations that represent the public health community. This includes work done cooperatively with national organizations like the Association of Public Health Laboratories (APHL), the Association of State and Territorial Health Officials (ASTHO), the National Association of City and County Health Officers (NACCHO), and the American Public Health Association (APHA). These organizations are an important part of BPRP's strategy for building preparedness for a number of reasons:

1. They provide CDC with a large pool of expertise and knowledge about the practice of public health at the state and local level;
2. They provide CDC with a feedback channel on the process of grant implementation and on the types of improvements taking place because of the grants;
3. They have been a source of information regarding the current status of the system and provide the institutional memory of how pre-BPRP efforts to improve public health have fared.

Early in the execution of the bioterrorism preparedness program CDC recognized the value of these organizations and worked to develop strong partnerships with them. CDC and APHL have developed a good relationship in developing the Laboratory Response Network, and both ASTHO and NACCHO have worked in closely in developing and administering the public health portion of the Department of Justice's vulnerability and needs assessment. In addition, CDC, through the Public Health Practice Program Office, has supported an exemplars program in which three local public health departments are working closely with NACCHO to continually assess the execution of their CDC bioterrorism grants and the progress in their functional improvements. Continued partnership with these organizations will only benefit preparedness efforts.

Many local public health officials have expressed a need for a "*national clearinghouse*" for bioterrorism-related issues, a focal point to receive questions and comments, and provide information and guidance, or a hub through which local entities can communicate across jurisdictions about preparedness-building experiences. As one local public health official put it, "On a number of occasions when I have had questions about surveillance initiatives or prophylaxis issues I called Annie Fein in the New York City Department of Health. She has been dealing with these issues for years and has been very helpful by sharing her wealth of knowledge, but she is also very busy and cannot be expected to answer my questions every time they arise."

BPRP should perform this "*clearinghouse*" function for information on activities of the states and localities dealing with the public health and medical dimension of bioterrorism preparedness and response. At the present time, however, it has little time for providing this service. With a larger staff, BPRP could act as a *clearinghouse* for bioterrorism public health and medical related issues. This would make them more responsive to local needs. First, it would facilitate communication not only in the direction of BPRP to local entities, but provide a mechanism for BPRP to listen and respond to the concerns of local entities. Second, BPRP could provide an extremely valuable service by assembling information—MMRS plans, Bioterrorism Response Plans, laboratory protocols, and general lessons learned—from all those entities involved in bioterrorism-related public health and medical activities. Depositing this information a single place for distribution among many jurisdictions is a much more efficient and less confusing pathway for information flows than forcing local entities to connect to many, many different informational sources.

One mechanism for providing this function is through an expanded BPRP web page. At present, the BPRP web page provides very little substantive information for local public health and medical officials. Providing materials such as hospital BT facility templates, BT Response Plan templates, educational materials for primary care physicians, and general lessons learned from experiences in other locales when preparing for bioterrorism would be useful to officials currently in the midst of BT planning activities.

The BPRP web page should also be used as a central depository for all public health and medical informational materials. Protocols for laboratory technicians who analyze BT samples are currently provided on the APHL website, while pharmaceutical and epidemiological information is difficult to find anywhere. The BPRP website should become the definitive website on public health and medical related BT activities. This will reduce confusion about where to find BT materials and will ensure quality control of the materials that are being provided. BPRP is embarking on a concerted outreach and education program that is aimed at critical constituencies for building BT preparedness. These efforts should be strongly supported, as should its plan to raise awareness of public health and medical dimensions of the bioterrorism issue on Capitol Hill.

While many senators and representatives are increasingly aware that bioterrorism could represent a serious national security threat, few know the intricacies of this threat or of the response to it, especially the public health and medical responses. While Congress has many members who are well schooled in national security affairs, only a single member is a medical doctor and only a handful are well versed in public health issues. Some concern exists that bioterrorism is the "policy issue of the month" in Washington and funding for building bioterrorism preparedness will quickly dry up when some other "hot" issue appears. BPRP should develop and implement a strategic plan to educate lawmakers about the importance of both BT issues and public health more broadly to ensure that efforts that are currently underway will be sustained financially into the future. Aside from lawmakers on Capitol Hill, BPRP could target educational efforts at hospital administrators and training centers, such as the Noble PHA Training Hospital, who have largely escaped involvement in BT preparedness activities, the media, and international entities because BT may be a foreign threat or a domestic one.

6

Epidemiology

Whereas surveillance is an *awareness tool* used to monitor and accumulate data about disease outbreaks, epidemiology is an *assessment tool* used to interpret the raw data gathered through various surveillance sources. While some assessing takes place during the course of surveillance—a physician recognizes and reports an unusual case or a medical examiner reports an unexplained death—it is largely left up to the epidemiologists to interpret surveillance data and recognize the significance of unusual occurrences. In determining the exact nature of an outbreak or *bioterrorist event*, the epidemiologist will conduct an investigation by working with a variety of data to determine the source of an outbreak, mode of transmission, extent of exposure, and how the outbreak is likely to spread. He will also predict future patterns, and calibrate the necessary response. Based on this information, epidemiologists make recommendations for the appropriate public health measures needed to contain the outbreak. The epidemiologist also informs the medical community of the nature of the problem they confront so that they can devise the appropriate treatment. In order to achieve these objectives, epidemiologists perform the following activities: (i) interpreting surveillance data; (ii) conducting investigations; (iii) building case definitions; and (iv) ongoing monitoring.

Interpreting Surveillance Data

While epidemiologists will be able to draw some preliminary conclusions about an outbreak from even the most basic *surveillance data*, the sophistication of what epidemiologists will be able to deduce from the surveillance data will depend on the quality of that data as well as how quickly it is received. Data gathered from surveillance will largely inform the epidemiological investigation at the onset, and the more thorough this data is, the closer it will bring the epidemiologist to the source and nature of the outbreak. For example, data that indicates a sharp increase in hospital admissions, while a sign that something unusual may be occurring, will be far less useful to epidemiologists than data that indicates syndromic information distributed geographically over time.

The former set of data indicates little more than the presence of an emerging epidemic, while analysis of the latter set of data can tell an epidemiologist something about the characteristics of the disease of concern, and roughly when and where the exposure occurred. If *surveillance data* is insufficient (i.e. single source—unusual clinical reports from physicians), then epidemiologists will have to go through the painstaking, time-consuming, and labour-intensive process of interviewing patients to collect data that can reveal necessary clues about the outbreak. Local departments of *epidemiology*, however, rarely, if ever, have the personnel to conduct large numbers of interviews of this type within the short timeline of a bioterrorist event.

Therefore, should it become necessary to conduct case interviews, it is likely that local and state officials will call on the CDC's Epidemic Intelligence Service (EIS) to support their efforts. If a request is made, the EIS, as it has in the past, will need to mobilize rapidly in support of a local investigation. Interpretation of raw data compiled through various forms of surveillance is a critical part of detecting and appropriately responding to a *bioterrorist attack*. Although hospital personnel and physicians are given guidance on reportable disease outbreaks and are required to report unusual patient cases, and laboratories are required to report unusual cultures to the local department of health, the current system is largely manually based. If an unusual case presents itself at a hospital, the department of health is likely to be notified by phone.

Even in instances in which locales have made arrangements with hospitals or laboratories for regular reporting of admissions or culture characteristics, this data must first be manually entered into a system and then forwarded to the department of health for analysis. There are no automated surveillance systems that gather surveillance data and forward it to the department of health hourly, after each shift, or at the end of each day. In the event of a bioterrorist attack, epidemiologists will need access to real-time surveillance data to monitor how an attack is progressing and unfolding. This is a major challenge to utilizing epidemiology effectively in a bioterrorism attack. Even if real-time surveillance data is available and forwarded to a department of health, someone must continually monitor that data to ensure prompt analysis and response.

Certain data are signals of a suspicious outbreak that should prompt epidemiologists to involve the larger public health and medical communities and request rapid laboratory identification of the *etiologic agent*. Some of these signals are:

1. Identification of rare or unusual cases of disease;
2. A surge in the number of people with similar syndromes;
3. A surge in the number of unexplained deaths;
4. Higher than normal morbidity or mortality associated with a syndrome;
5. Multiple diseases existing in a single patient;
6. Higher than normal or unseasonable occurrence of a disease;
7. Unusual geographic distribution;
8. Atypical presentation of disease;
9. Endemic disease with an unexplained increase in incidence;
10. A genetically altered or atypical strain;
11. Atypical aerosol, food, or water borne transmission; and
12. Concurrent animal outbreak.

If signals of a suspicious outbreak are present, epidemiologists should alert public health and medical personnel as well as law enforcement to the emergence of a suspicious outbreak of disease so that a response can be initiated. Once the larger public health and medical communities have been alerted and advised to report unusual illnesses, identification of the etiologic agent by a laboratory must be initiated immediately to determine if criminal activity has occurred and to initiate the appropriate measures for containing and treating an outbreak.

Epidemiologists are unlikely to be able to make a positive confirmation until surveillance and patient data begins to accumulate several days into the event. Many bioterrorism agents, however, have very short *incubation periods*—3 to 5 days from time of exposure to becoming symptomatic—and treatment is often not effective after patients become *symptomatic*. Because these symptoms are non-specific and flu-like, physicians and epidemiologists may not recognize that something unusual is occurring until an unusual number of patients begin to present (unless there are obvious indications of bioterrorism or if the attack is announced). These time constraints could limit the initial role that epidemiology will play to that of analyzing data and determining the extent of the exposure and how the outbreak is likely to progress over time. If epidemiologists are to play a central role in detecting a bioterrorist attack in its earliest stages, they must be provided with quality, real time surveillance data, preferably distributed geographically so that epidemiologists can not only detect, but characterize the event and guide the necessary response.

Outbreak Investigation

Traditionally, *epidemiology* is a slow, labour-intensive process of investigating disease outbreaks. Epidemiology usually begins after patients have presented at hospitals or physicians offices with unusual symptoms. If background or baseline disease data suggests that patients are presenting at unusual rates or in unusually high numbers epidemiologists must initiate an investigation. This involves interviewing patients to determine life-style patterns and identify commonalities between patients that may indicate the source of the outbreak and mode of transmission so that public health measures can be taken to contain the outbreak. If the source of the outbreak cannot be determined from patient interviews, epidemiologists will collect and test environmental samples—food or water samples, insect and animal vectors, or soil samples—to identify the source of the outbreak and mode of transmission.

While traditional epidemiology is useful for understanding and containing slow moving disease outbreaks, an epidemiological investigation will need to move much more rapidly if a bioterrorist attack is suspected. Although the ability of epidemiology to guide the bioterrorism response system in the early hours and days of a bioterrorist attack will be limited because of the pace at which events will unfold, epidemiology will be important as a bioterrorist attack plays out over time, especially if non-traditional BW agents are utilized.

The crucial moments of recognizing a bioterrorist attack are likely to unfold over a period of days, but the effects are likely to continue over a period of weeks. For instance, after the Sverdlovsk *anthrax* release, approximately one-third of the patients presented with symptoms in the first week, another one-third presented in the second week, and the final one-third presented sporadically over the next 3 to 6 weeks. The Sverdlovsk experience is telling given that a bioterrorist attack is likely to involve less than optimal (from the attacker's viewpoint) environmental conditions, agent quality, and dissemination techniques, and thus produce less than optimal results. As a bioterrorism scenario plays out over several days to weeks, and as patient data begins to accumulate, epidemiologists will have a central role in analyzing that data to determine:

1. The approximate point of exposure and the population most likely to have been exposed so that prophylaxis and treatment can be focused there first;
2. Measures for containing the outbreak;
3. Whether a single attack or multiple attacks occurred;
4. Whether follow-on attacks have been carried out that may result in additional waves of patients;
5. How the outbreak will unfold over time; and
6. Clues that may aid a law enforcement investigation.

Environmental samples are important in identifying the source and nature of an outbreak. If an approximate release location can be determined from patient interviews, then analyses of soil samples may indicate the presence of a non-endemic microbe or a higher than normal presence of a particular microbe. The presence of an unusual number of dead animals or the presence of abnormal symptoms in animals treated by veterinarians may prove to be a key to understanding an outbreak. The West Nile outbreak in New York City is a good example; dead crows became central to discovering that the outbreak was caused by West Nile virus and not St. Louis encephalitis.

Epidemiologists will collect environmental samples such as rodents, insects or other animal vectors for laboratory analysis to determine the reservoir of the outbreak. Good surveillance data accumulated by veterinarians will prove very helpful to epidemiologists, especially when zoonotic diseases are involved. Epidemiologists will assess the type of outbreak as well as advise on the type of measures that should be taken to contain the outbreak. Containment of a *contagious disease* will require extremely effective disease control and quarantine measures.

Epidemiology will be important for monitoring patient data and animal illnesses to determine the approximate point of exposure so that *prophylaxis* or antibiotic treatments can be focused on people who can be placed in that particular geographic area at the time of the attack. Epidemiology will also be important for monitoring and securing a contaminated water or food supply, and infection control measures will depend largely on the information gathered during an epidemiological investigation. Epidemiology will also play an important role in *bioterrorism events* that do not involve agents of highest concern. Because inhalation *anthrax* or *plague* is almost non-existent, a single case is enough to suspect bioterrorism. But in many other deliberately induced disease outbreaks, linking the disease to bioterrorism may not be obvious. A *bioterrorism attack* may not necessarily have a steep epidemic curve. In the case of a food-borne attack in which the source of the outbreak is a food product that is purchased and consumed gradually over the period of days to weeks, the contaminated products would result in a slow rise in an epidemic curve that may resemble natural phenomena.

Small-scale food and waterborne attacks are easier to conduct than those using aerosols, and it is difficult to distinguish a natural outbreak from a bioterrorism event. It was thought that the Rajneeshee salmonella poisoning in The Dalles, Oregon was caused by poor sanitary conditions at restaurant salad bars and was only discovered to be the result of bioterrorism over a year later, and only then after a cult member confessed to the deliberate contamination. In food and waterborne incidents, epidemiologists will have an important role in determining the source of the outbreak so that access to and distribution of the materials are halted.

Constructing Case Definitions

Even with relatively little *surveillance data*, epidemiologists can create a case definition from the beginning of an outbreak in order to alert public health and medical personnel of the need to report cases that are similar to the case definition and request laboratory cultures to confirm that a patient has been afflicted. A case definition will initially describe known symptoms: fever, chills, nausea, headache, vomiting, etc., the geographic location of patient clusters, and a time window of exposure (if known). It will also provide physicians with treatment protocols and advanced clinical symptoms. A case definition is an important tool for aiding physicians and public health officials in identifying, reporting, monitoring, and tracking both old and new cases related to the attack.

Construction of a case definition will be essential to medical personnel tasked with triage responsibilities as guidance for distinguishing between potential attack victims and the "*worried well*," and directing these individuals to the appropriate treatment regimen. The starting point for constructing a case definition will be surveillance data. Epidemiologists will then begin to fill in gaps in that data by compiling additional information. Epidemiological investigations will include interviews with patients to uncover common linkages and identify additional people who may have been exposed. If a common facility is linked to a high percentage of case patients, the administrative records of that facility—work schedules or time cards, visitor sign-in sheets, personal contact information on employees—may be reviewed to identify potential victims. Press releases, news stories, advertisements, or other communications means may be used to encourage individuals that have information pertaining to the outbreak to come forward or to direct individuals with symptoms similar to those described by the case definition to the appropriate treatment center where additional information can be gathered during the course of treatment.

Monitoring a Bioterrorism Event

Simple case accumulation will allow epidemiologists to calculate an epidemic curve and give a rough estimation of when the initial exposure occurred based on the average incubation period for the particular disease. But this calculation will depend on positive laboratory confirmation of the etiologic agent. The pattern of the epidemic curve will also be important for determining the nature of an outbreak, and may, but will not always, signal that an outbreak is intentionally induced. A rapid spike in the epidemic curve may indicate that a *bioterrorist attack* has occurred if exposed victims become ill at approximately the same time after exposure. Most natural outbreaks gradually accumulate as people come into contact with the source of the outbreak over time, although sharp spikes are sometimes seen in natural point source exposures.

Eventually all of those people who are likely to come into contact with the source of a natural outbreak will have done so and the curve begins to tail off, but in the early stages, a natural and intentional outbreak spike may be indistinguishable. This is especially true if the outbreak presents itself over a longer period of time. Epidemiologists will continue to gather information from the Office of the Medical Examiner, local coroners, veterinarians, toxicologists, medical providers, and hospitals, and analyze this information to make judgments and projections about the nature and course of the outbreak. Once an outbreak is underway, epidemiologists need continued access to information on patient load and symptoms, as well as laboratory results, and they must have a means of providing information back to public health and medical officials who can use it to make critical decisions about treatment of patients. A distilled version of the epidemiological data will need to be disseminated to the response entities to keep them abreast of developments, and some information must be provided to the public about the nature of the outbreak and practices and procedures for containing it.

At present there is no mechanism for epidemiologists and other public health and medical entities to receive or exchange surveillance data or add to that data via a central information and communications system. In summary, while the practice of *epidemiology* is very similar in natural outbreaks and *bioterrorist attacks*, there is a basic set of requirements that are helpful in conducting any epidemiological investigation, and other requirements that can greatly benefit epidemiology during a bioterrorism event. These requirements include:

1. Baseline disease data;
2. Real-time access to surveillance data;
3. Adequate personnel to continuously analyze surveillance data and investigate unusual outbreaks;
4. Personnel and equipment to gather environmental samples;
5. Robust laboratory capacity to test environmental samples, confirm epidemiological findings, and provide additional information about the demographics of patients;
6. IT systems that can compile and analyze patient data gathered manually during epidemiological interviews;
7. Ability to communicate easily and share information with laboratories, hospitals and physicians and federal level entities;
8. Broad understanding of a variety of disease patterns—endemic, non-endemic, food and waterborne, traditional bioterrorism agents, as well as unexpected or non-traditional bioterrorism agents.

In essence, epidemiology engages in the process of collecting disease data using a variety of tools, analyzing that data, and making recommendations for controlling the outbreak and maintaining public health. Many of these requirements are labour-intensive and require staff to meet them, yet building epidemiological capacity at the local level is a daunting challenge.

Building Epidemiological Capacity

Because surveillance begins at the local level, and because detection and response to a *bioterrorism attack*—at least initially—will most likely occur on the local level, epidemiological capacity must begin with local public health departments. In states with largely rural counties that do not have the resources to maintain a staff of epidemiologists or even a single dedicated epidemiologist in some cases, this capacity should be built primarily at the state level. However, it should be responsive enough that the capability can be mobilized to support an investigation at the county level in the event of a bioterrorist attack or disease outbreak. To build this capacity from the ground up in rural states is a daunting task that would require enormous sums of money; anything short of significant sums would result in uneven and sporadic epidemiological coverage throughout the state. Such expenditures must be balanced by the reality that *bioterrorism attacks* in rural areas are in and of themselves of lower probability than attacks in urban areas and far less likely to cause overwhelming casualties because of lower population densities.

Capacity must also be built at the city level in larger urban areas where disease outbreaks are more likely to be significant and bioterrorism attacks are more probable. States and regions with several large cities in fairly close proximity should build a robust epidemiological capacity at both the state and local levels. Large urban centers already supply epidemiologists with numerous challenges in containing disease outbreaks and improving public health.

Bioterrorism is one more challenge that epidemiologists in urban centers must face. While epidemiologists in a particular city may be able to monitor the health of that city, urban populations tend to move significantly on a daily basis. In the event of a bioterrorist attack, a city epidemiologist will not be able to track potentially exposed victims who have left the city for the suburbs or another city in that state or elsewhere. For this reason, it is essential that epidemiological capacity exists on the state level in regions with multiple large urban centers. For example, victims from a bioterrorist attack in New York City are likely to present in the city, but also in New Jersey, Connecticut, upstate New York, and possibly even Boston, Philadelphia, or Washington. Therefore, epidemiological capacity will be required on the state level to investigate the relationship between what is occurring as the result of a bioterrorism attack in New York City and syndromic trends in a broader geographic area.

Where Are We?

Epidemiological capacity at the local level is currently lacking. Many state and most county health departments lack the funds to add staff to address bioterrorism issues. Instead, responsibilities for building epidemiological capacity for responding to bioterrorism are being handled by current staff members whose time is already consumed by investigations of natural disease outbreaks—*influenza*, food-borne illnesses, sexually transmitted diseases—or by public health campaigns aimed at generally improving the health of local populations. This approach leaves only a small percentage of time, if any, for bioterrorism-related planning. Public health officials in New York City trying to cope with West Nile virus, for example, have little if any time to deal with normal problems, let alone bioterrorism preparedness efforts. While some health departments have received federal grant money for bioterrorism planning activities, and have been able to use these grants to supplement their administrative staff or bring on an additional epidemiologist, most health departments have not received planning grants for these purposes. The result is added workloads for current staff members.

Even when health departments are engaged in planning and preparedness activities, because epidemiological investigations are labour-intensive, there is still insufficient local capacity for conducting rapid and wide-reaching investigations in the event of a bioterrorist attack. In many cases, then, epidemiological support will have to come from state health departments where available, or rely on the CDC's epidemiological assets. Although CDC's epidemiological assets are exceptional, a major bioterrorist attack would seriously tax these resources, leaving little excess capacity to deal with multiple attacks in different locales or other health concerns.

Local health departments also do not have the capacity to dedicate a staff member to developing relationships with potential providers of *surveillance data*—hospitals, primary care physicians, laboratories, pharmacies, school and employment personnel offices, etc.—or devising ways to implement and installing data gathering or communications mechanism with these entities. This is partly due to a lack of staffing but derives from the lack in local health departments of *information technology* (IT) to gather *surveillance data* in real time, share that information, or communicate with response entities during an event through a common IT system. This is partly due to a lack of funding for putting IT infrastructure into place.

Surveillance and Monitoring

Without adequate *surveillance* and *monitoring*, it is unlikely that epidemiologists will detect a bioterrorist attack early on. Rather, it is more likely that an attack will be detected after the event when a surge of symptomatic patients begins to present at emergency rooms and doctors offices. At this point it may be too late to effectively treat the first wave of victims. Without adequate surveillance and monitoring, it will also be extremely difficult for epidemiologists to determine the nature of the event. Because epidemiologists require quality data to do their job, it is critical to develop better disease surveillance systems and bolster epidemiological staff in state health departments.

Epidemiological Capacity

Current *epidemiological capacity* in local or state health departments is lacking and would be overwhelmed by a *bioterrorist attack*. This is largely because epidemiology at the local level has been neglected over the past two decades. Rebuilding this capacity is an enormous task. *Epidemiology* is labour intensive and staff is required to establish agreements with the providers of *surveillance data*, monitor, and analyze this data in real time, conduct patient interviews, and collect environmental samples. Local health departments do not have the staff to perform all of these functions under normal operating conditions, let alone under crisis conditions characteristic of a bioterrorist attack.

The labour-intensive nature of epidemiology requires that more funding be provided to state departments of health to hire additional epidemiological staff. This should be a priority focus area. The objective, however, should not be only to build assessment tools that make the response to bioterrorism more focused and efficient. It should also have practical benefits for responding to natural outbreaks of disease or for mitigating public health threats generally. In rural states like New Mexico or Montana, building epidemiological capacity at the local level is prohibitively expensive and largely unnecessary. This capacity should be built at the state level. In mixed urban and rural states such as New York or Illinois, urban centers should maintain a robust epidemiological capacity to deal with the myriad of public problems in that city, including bioterrorism, but efforts must be balanced with state efforts to develop statewide epidemiological capabilities.

Response Triggers

A tension exists between the need to initiate treatment and *prophylaxis* as quickly as possible and the need to know the nature and extent of a *bioterrorist attack* before initiating a massive response. Thresholds for triggering particular responses should be defined to avoid unnecessary "*hair trigger*" responses. The first inclination of a suspicious outbreak should trigger an *initial response phase*—alert hospitals and physicians, require doctors to take culture samples, seek laboratory diagnosis, and notify the appropriate federal, state, and local authorities. *Large-scale mobilization* of the response system should require that a second threshold be crossed, for example, laboratory identification of the etiologic agent representing an imminent public health threat or the accumulation of surveillance data that indicates such measures are necessary.

Diverse Knowledge of Potential Bioterrorism Agents

A range of possible agents may be used in a bioterrorist attack to varying degrees of effectiveness. CDC's bioterrorism agents list encompasses a broad spectrum of possible agents, and considerable work is being done on many of them. Much of the focus has been on high impact agents such as *smallpox*, *anthrax*, *plague*, and *tularemia*, which are usually associated with mass casualties. This has created an assumption that if bioterrorism occurs it will involve one of these agents. Epidemiologists need to be made more aware of the full range of potential bioterrorism agents, not just those few that have been the focal point of most discussions to date. CDC should develop a mechanism, therefore, that conveys new developments to epidemiologists as its work generates new information regarding the full range of agents. Although primary attention must continue to be paid to agents that are most likely to be used or have the greatest potential impact in terms of casualties, periodic seminars or other awareness-raising mechanisms should be developed to inform state and local epidemiologists about those agents that are not at the top of the CDC list.

Information Technology

Local departments of health need funds to make use of, and integrate, information technologies that can be used to collect, deposit, analyze, and share surveillance and epidemiological data from a central location in the department of health. Because the HAN network underpins many of the bioterrorism assessment and response tools, including epidemiology, it is essential that this infrastructure be expanded in ways that facilitate other tools.

7

Emerging Diseases

Since the end of the *Cold War*, the foreign policy community has debated the role of environmental issues in *national security*. Just as the oil shocks of the 1970s demonstrated that traditional defense-oriented notions of security had to be broadened to include economic issues, so some now argue that certain environmental issues should be included as well. But not all environmental threats may be cogently and usefully framed as security issues. Such threats must be sufficiently understood scientifically for proposed solutions to be credible, and they also must be of enough potential consequence to justify the commitment of resources and the attention of senior policy makers. Few environmental threats have been perceived across the political spectrum as satisfying both these requirements, and few have achieved broad recognition as security issues. Emerging infectious disease may provide the strongest example of an environmental challenge that poses a clear threat to national security.

Emerging infections undermine the societies of the most disease-stricken countries, readily cross international borders, and consequently directly threaten the health of people throughout the world. The 1918-19 Spanish influenza pandemic killed twenty to twenty-five million people worldwide, including one-half million Americans. The *human immunodeficiency virus* (HIV), which infects over percent of the population of sub-Saharan Africa, remains one of the ten leading causes of death in the United States. Even definitions of national security restricted to defense and economics alone must acknowledge the relevance of epidemics of this magnitude and the importance of preventing future outbreaks. Moreover, there is the possibility that a major outbreak of disease could originate not from natural causes, or unintended consequences of human behaviour, but rather as a result of deliberate biological attack. Diseases that enter the human population from some natural reservoir are said to have "*emerged.*"

Diseases emerge for many, often poorly understood, reasons. Mad cow disease most likely emerged when humans ate infected cattle products. The horrific *Ebola virus* lives in an unidentified animal or insect host and only now and then jumps to humans, for reasons that remain largely unknown. The fatal hantavirus outbreak in the American Southwest in 1993 was traced back to deer mice whose population had exploded due to local climate change: food sources had been unusually abundant that year because of record precipitation. In other cases, such as that of tuberculosis in the United States, diseases once thought to be under control developed resistance to known antibiotics and experienced a resurgence.

Epidemic diphtheria returned to Eastern Europe as vaccination programs weakened, the pool of immunologically compromised individuals in major cities expanded, and infected travelers spread the disease. Even excluding HIV/AIDS (itself a recently emerged disease), the death rate from infectious diseases rose by percent in the United States between 1980 and 1992. It is anticipated that as more organisms develop resistance to over-prescribed antibiotics, as the growing human population increases its inroads into tropical forests, as cities in the developing world become more crowded and sanitation ever more problematic, and as greater numbers of people travel or flee across international borders, more "new" diseases will appear or take hold in the human population.

Threat of Biological Terrorism

As was true with emerging disease, the threat of biological terrorism was present throughout the *Cold War*, but concern over this threat has since escalated. During the Cold War, the greatest biological threat faced by the United States was posed by the Soviet Union, and that threat seemed most likely to be manifested in the event of overt war rather than in acts of terrorism. Nations currently regarded as rogue states were less likely during the Cold War to have the freedom of action to launch terrorist attacks using weapons of mass destruction. With the collapse of the Soviet Union and the current overwhelming conventional military superiority of the United States, weapons of mass destruction might now be seen by these nations as equalizers, to be brandished or used in terrorism or unconventional warfare.

Moreover, the human and technical resources needed for a biological weapons program are spreading throughout the world. While it may still be true that most developing countries lack the microbiological capacity needed to develop biological warfare agents, eight or more developing nations have been implicated in developing an offensive biological warfare capability. It has been alleged by a prominent Russian defector that Russian scientists have used genetic engineering to produce antibiotic-resistant strains of a number of disease organisms. Some of this research has been published in the open literature, so it is available for application by nations developing their own biotechnology industries. Worse yet is the possibility that some Russian researchers may market their personal expertise to other nations. The world recently learned of the Aum Shinrikyo, a multinational terrorist group intent on the development and use of an array of weapons of mass destruction.

Biological and chemical agents are now demonstrably within the technical expertise of such groups, and even primitive and inefficient versions of these weapons could have devastating economic or terror impacts. With the independent bombing attacks at the World Trade Center in New York in and the Federal Building in Oklahoma City in , the release of sarin nerve agent in the Tokyo subway system in by the Aum Shinrikyo, and that group's attempts to attack Tokyo using anthrax and botulism, large-scale terrorist attacks on civilian populations using weapons of mass destruction no longer seem in the realm of the fantastic. At their worst, the New York, Oklahoma City, and Tokyo attacks may represent the crossing of a grim threshold, weakening long-standing taboos and increasing the likelihood of analogous attacks in the future.

Bioterrorism: A Unique Threat?

Biological weapons differ fundamentally from other *weapons of mass destruction*. Whereas nuclear and chemical weapons cause immediate casualties, biological agents require hours to days or even weeks of incubation before they cause fatalities. (Exposure to some biologically derived toxins, as opposed to living organisms which incubate inside the body, could exhibit timescales more similar to those of chemical weapons—but *toxins* are, in effect, biologically derived chemical weapons.) Barring a terrorist announcement or fortuitous discovery, a biological attack will first become known hours or days after its execution, when its victims begin to appear at doctors' offices and hospital emergency rooms. Sufficiently subtle biological terrorist attacks might go unrecognized, or remain undetected, for long periods of time. Terrorism involving the dispersal of radiological materials could, in this respect, share some similarities with biological agents.

Desirable characteristics for biological agents for military use include minimal contagiousness (as with *Bacillus anthracis*, the sporulating bacterium that causes *anthrax*) to ensure that the disease cannot produce an uncontrolled epidemic that could boomerang and infect the attacker's forces or population. For a terrorist attack, however, contagiousness (communicability) might well be viewed as an asset. *Contagious agents* may spread the disease far beyond the initially exposed population. For example, the two- to three-day incubation period for plague (*Yersinia pestis*) is long enough to allow victims of an attack to travel by air between virtually any city in the world and any city in the United States before falling seriously ill—especially so if the organisms were released in the departure gates area of a major international airport.

The *smallpox virus* has an incubation time of seven to seventeen days. The discontinuation of routine vaccination, the contagiousness of the disease (including secondary transmission from infected individuals who might themselves never manifest the disease), and the unknown extent of possible clandestine stockpiles make smallpox seem an especially attractive agent of *bioterrorism*. It has been reported that North Korea may retain smallpox cultures for use as a biological weapon.

Consider the outcome of one infected individual visiting New York City. In 1947 an American businessman arrived from Mexico with fever, headache, and rash, and then spent several hours sightseeing. His illness turned out to be smallpox, and he died nine days later after having infected twelve others, of whom two died. Public health officials viewed the potential for transmission to be so serious (despite the fact that Americans at that time were routinely vaccinated against smallpox) that over six million people in New York City were vaccinated within a month. In contrast to 1947, Americans have not been routinely vaccinated against smallpox since 1980. The federal Centers for Disease Control and Prevention (CDC) currently maintains over twelve million doses of vaccine in storage.

In a more recent case, in 1972 a pilgrim returned to Yugoslavia from Iraq infected with smallpox. To contain the resulting outbreak, over ten thousand people were quarantined, and twenty million were vaccinated in less than two weeks. The incubation delay endows biological agents with advantages as terrorist weapons that nuclear or chemical weapons lack. The possibility of contagion (in the case of some agents such as smallpox), and resulting fear, is a further terror advantage. These aspects of biological agents emphasize the importance of *a strategy of public health surveillance* for incidents of bioterrorism—a strategy that is inapplicable to the cases of chemical or nuclear attacks. The threats posed by biological weapons must be thought about and addressed in a very different way.

Barring a terrorist announcement or an interruption of an attack while underway, traditional "*first responders*" (fire, police, paramedics) or quick-response teams with specialized equipment and training will not be among those initially recognizing and responding to a biological attack. As described below, biological terrorist attacks conducted by followers of Baghwan Shree Rajneesh in 1984 and attempted by members of the Aum Shinrikyo in the early s were accompanied by no announcements and remained unrecognized while in progress. As these examples illustrate, recognition and response to a successful biological attack, domestically or abroad, will most likely depend upon the sensitivity and connectivity of the existing public health system.

Sensitivity refers to how likely it is that a given presentation of a disease will be recognized by a physician or other health-care worker as being out of the ordinary. *Connectivity* refers to how quickly and accurately information about a case gets passed "*vertically*" from the clinical level up to state, national, or international authorities, and "horizontally" within these levels.

Within the United States, surveillance for infectious diseases is a largely passive process. Each state has its own requirements for the reporting of specific diseases by physicians, hospitals, and other health care providers. Local or state health departments are supposed to be notified by physicians or laboratories if a patient is diagnosed with a disease defined as reportable by that state. At the national level, the CDC collaborates with the Council of State and Territorial Epidemiologists in maintaining a list of national notifiable diseases. Individual states' requirements for notifiable diseases typically parallel this list. Through the National Notifiable Diseases Surveillance System (NNDSS), states voluntarily report weekly to the CDC on the incidence of some fifty diseases. These diseases include several of potential interest to bioterrorists, such as anthrax, botulism, brucellosis, and plague. Reporting is mandatory for a small number of diseases requiring quarantine, such as suspected smallpox, infectious tuberculosis, and viral hemorrhagic fevers. Outbreaks of diseases not on the national notifiable diseases list may remain undetected until an outbreak is well under way. The CDC analyzes the data it receives and reports on it in its *Morbidity and Mortality Weekly Report*.

A second type of national disease surveillance by the CDC involves the use of *"sentinel" hospitals*. In this case, no attempt is made to gather comprehensive national data. Rather, the National Nosocomial Infection Surveillance (NNIS) system gathers data voluntarily provided by 163 (as of 1993) hospitals. (Nosocomial infections are infections acquired while a patient is hospitalized.) Incidence of infections in the participating hospitals may be used to estimate the national incidence of nosocomial infections. In addition to the NNDSS and the NNIS system, the CDC also engages in pilot projects with certain states, in key U.S. cities, or with individual "*sentinel*" physicians, to gather data on the incidence and characteristics of other diseases or disease outbreaks.

A survey of the possible scales of terrorist attacks and the extent to which these have proven or may prove difficult to distinguish from outbreaks of infectious diseases, will make it clear that improving surveillance for biological terrorism must build directly upon existing public health surveillance systems. As discussed below, this requirement has specific implications for the steps that should be taken to prepare for *bioterrorism*.

Possible Scales of Terrorist Attacks

An examination of *biological terrorism* episodes illustrates the range of threats for which we must prepare. The most dramatic *biological threat* is a major terrorist attack against an urban center, using an efficient mechanism for the dispersal of the biological agent. In 1993, the *Office of Technology Assessment* (OTA) estimated that kg of aerosolized (converted to respirable particles in the 1 to 5 micron size range) *Bacillus anthracis* spores dispensed by an airplane upwind of a major city could kill hundreds of thousands to millions of people. A different scenario estimates that the number of deaths resulting from an anthrax aerosol dispersed from a boat sailing upwind from New York City could be over 400,000, people. While no such major biological attack has yet succeeded, this decade has seen the release of sarin nerve agent in the Tokyo metro system in 1995 by the Aum Shinrikyo religious cult (killing eleven people, with over 5,000 injured, of whom some 700 required hospitalization). Moreover, the Aum repeatedly—at least nine times—attempted biological attacks on Tokyo city as well as nearby U.S. naval installations. While the failure of the Aum's attacks suggests that acquiring and successfully weaponizing an effective biological agent remains challenging, large-scale attacks on civilian urban populations nevertheless are clearly no longer in the realm of the fantastic.

A large-scale attack against an urban center using biological agents would be the manifestation of *biological terrorism* having the most in common with a chemical or nuclear attack. Even for a massive urban biological attack, however, public health surveillance could be critical to minimizing deaths and casualties, as well as economic costs. A recent study examined expected deaths and economic impact for scenarios involving three different biological agents (*Bacillus anthracis*, *Brucella melitensis*, and *Francisella tularensis*) released as aerosols in a terrorist attack on a major city. The timescales required for effective intervention vary according to the agent. First, consider the anthrax case as an example. The study found that intervention (taken to be 90 percent effective administration of antibiotics and vaccinations) within one day after the attack could keep deaths to below 10,000, as opposed to over 30,000 if intervention occurred five or more days later, and could save $15 billion to $20 billion. At the other extreme in incubation timescales, for the case of brucellosis, intervention within the first two weeks after the attack would reduce deaths to about 100 compared with over 500 if intervention did not take place until after two months. Effective intervention would only be possible if family physicians and emergency room personnel recognized as early as possible during an outbreak that an anomalous situation existed, and if these concerns were effectively passed to state and national health authorities for rapid diagnosis and response.

These same public health capabilities are those necessary to detect more subtle attacks as well. One can envision terrorists introducing a disease into the United States in such a way that no easily recognizable outbreak occurs, or so that no outbreak is noticed until the disease is well under way. Such a masked attack, followed by a credible terrorist announcement, could have an impact far out of proportion to the deaths that actually resulted.

It has sometimes been asserted that there is an outstanding puzzle (perhaps of a psychological nature) to be solved regarding why terrorists have so far been deterred from biological attacks. In light of the attempts by the Aum Shinrikyo to attack Tokyo with biological weapons, this supposed puzzle now seems moot. Indeed, biological terrorist attacks have been conducted or attempted at many scales. The following examples illustrate this point and demonstrate the range of incidents over which surveillance must be effective.

Individuals

In two members of the Minnesota Patriots Council were convicted of planning to use the biological toxin ricin to assassinate Internal Revenue Service agents and a deputy U.S. marshall. In that same year, a member of the white supremacist organization Aryan Nation was arrested for ordering three vials of freezedried bubonic plague from American Type Culture Collection, a biological supply house in Rockville, Maryland.

Assassination Campaigns

Testimony in June before South Africa's Truth and Reconciliation Commission revealed that the apartheid-era South African government developed chocolates and cigarettes infected with anthrax, beer bottles containing botulism, sugar laced with *Salmonella*, and bottles of cholera culture. These products were used both for the attempted assassination of specific political opponents and, perhaps, to cause outbreaks in African National Congress training camps.

Workplaces

In 1996, twelve laboratory workers at a large medical center in Texas developed acute diarrheal illness after eating doughnuts left in their break room that had been intentionally contaminated with the bacterium *Shigella dysenteriae*.

Local Communities

In September 1984, members of an Oregon commune headed by the Bhagwan Shree Rajneesh used *Salmonella* to contaminate restaurant salad bars and coffee creamers in The Dalles, the county seat of Wasco County, Oregon. Although there were no fatalities, some 750 people became ill, with 45 requiring hospitalization. The outbreak strain of *Salmonella typhimurium* was shown in 1985 to be the same as a culture of *S typhimurium* found by an Oregon Public Health Laboratory official in a clinical laboratory operated by the commune. Two commune members were indicted in 1986 and later pleaded guilty to conspiring to tamper with consumer products by poisoning food. The commune members had been testing a plan to incapacitate voters in preparation for an upcoming election, intending to influence the outcome by making citizens of The Dalles sick on election day. Public health authorities had initially rejected the possibility of intentional contamination in this case, in part because they assumed that terrorists would issue a public statement in order to create widespread fear, rather than engage in a covert attack.

Cities—Water Supplies

In 1972, members of a U.S. fascist group called the Order of the Rising Sun were arrested in possession of 30 to 40 kg of typhoid bacteria with which they planned to contaminate water supplies in Chicago, St. Louis, and other midwestern U.S. cities. It is unlikely such an attack could have been successful, due to chlorination.

Cities—Aerosol Release

In the early 1990s, the Aum Shinrikyo released anthrax bacteria at least twice from a building in eastern Tokyo. Similarly, they sprayed anthrax in aerosol form from a truck driven around Tokyo, and they released botulism in a similar manner. None of these attacks appears to have resulted in any casualties, and it appears that the Aum both failed to breed the most virulent strains and did not master aerosolization.

Attacks on U.S. Military Bases

The Aum Shinrikyo reportedly also sprayed anthrax from a truck driven past the U.S. naval installation at Yokohama, then by the headquarters of the U.S. Navy's Seventh Fleet at Yokosuka. Again, neither of these attacks appears to have resulted in any casualties.

Accidental Releases from Biological Warfare Facilities in the Soviet Union and the United States

In , an unusual anthrax epidemic occurred in the Soviet city of Sverdlovsk in the former U.S.S.R. Soviet officials attributed the outbreak to consumption of contaminated meat, whereas U.S. agencies suspected it to be due to the accidental release of spores from a military facility located in the city. In 1992 Russian President Boris Yeltsin, who in 1979 had been the chief Communist Party official of the Sverdlovsk region, stated that "the KGB admitted that our military developments were the cause." Subsequent analysis of epidemiological data confirmed that the pathogen had been airborne, and allowed the location and date of escape to be identified.

The latter example suggests that international surveillance for biological terrorism may encounter examples of accidental as well as intentional release. This possibility is supported by the history of the U.S. biological weapons program. In the U.S., an offensive biological weapons program was begun in 1942 with research and development facilities at Camp (later Fort) Detrick, Maryland, testing sites in

Mississippi and Utah, and a production facility in Terre Haute, Indiana. This production facility lacked adequate engineering safety measures, and tests of the fermentation and storage processes using nonpathogenic bacteria demonstrated contamination of the plant and its environs.

Inspections of biological facilities in Iraq by the United Nations Special Commission (UNSCOM) indicate that the Baghdad government cut corners on safety and biocontainment, viewing production workers as expendable. Some governments may be especially likely to infect their own citizens during weapons development and production. These examples suggest that unintentional releases of biological agents may be typical events in a developing biological weapons program. International surveillance capable of investigating accidental releases is therefore important.

Reservoirs of Ambiguity

Analysis of clinical samples and/or epidemiological data may allow a distinction to be made between naturally occurring illness and intentional attack; modern DNA sequencing techniques should enhance this capability. From the public health standpoint, whether an outbreak is natural or artificial may be of little significance, though the political or legal ramifications of that distinction could be large. The inextricable relationship between surveillance for biological terrorism and surveillance for naturally occurring diseases becomes clear through the consideration of further examples illustrating the potential difficulty in disentangling natural outbreaks from certain biological attacks.

Legionnaire's Disease

In 1976, 221 people suddenly contracted pneumonia at an American Legion convention in Philadelphia, leading to thirty-four deaths. The bacterium *Legionella pneumophila* had multiplied in the water tower for the hotel's evaporative cooling system, exposing many guests to an infective dose of the organism. It was later shown that sporadic illnesses in 1947 and an outbreak in 1957 had in fact been due to "*Legionnaire's Disease*" but had not been identified at those times. In both the case of Legionnaire's Disease and in the outbreak of Hantavirus pulmonary syndrome in the southwestern U.S. in 1993 (fatal to 50 percent of those infected), there was initial concern that the illnesses might be due to criminal or *terrorist attacks*.

Plague in Surat

The outbreak of pneumonic plague in Surat, India, was depicted by a cover story in the Indian national newsweekly *The Week* as being due to biological warfare experiments conducted by the United States. While scientific refutations are unlikely to be entertained by the most extreme of those making such accusations, this incident, along with the previously cited domestic U.S. examples, serves to emphasize the importance of determining that certain outbreaks are *not* the results of terrorism.

Intentional Introduction of Diseases using Natural Vectors

During the Second World War, the Japanese attacked at least eleven Chinese cities with biological agents, contaminating food and water supplies with *Bacillus anthracis*, *Vibrio cholerae*, *Shigella*, *Salmonella*, and *Yersinia pestis*. Plague was developed as a weapon by allowing fleas to feed on plague-infested rats; as many as 15 million fleas were then harvested and released from airplanes per attack over Chinese cities. More recently, in 1969 the medical attaché to the French Department of Overseas Territories was quoted as saying that in Brazil, infectious organisms "were deliberately brought into Indian territories by landowners and speculators utilizing a mestizo previously infected " leading rapidly to the deaths of many Indians, who lacked immunity. Between 1957 and 1963, the attaché said, outsiders intentionally introduced smallpox, influenza, tuberculosis, and measles to the tribes of the Mato Grosso region. In 1964 and 1965, tuberculosis was allegedly intentionally introduced into the northern section of the Amazon Basin.

Aum Shinrikyo and Ebola

In 1992, Shoko Asahara, the head of the Aum Shinrikyo, and some forty followers traveled to Zaire, evidently with the intention of obtaining samples of the Ebola virus to culture and use in biological attacks.

Foodborne Outbreaks

Diseases borne by domestic and foreign foods kill 9,000 Americans each year, and sicken millions. Almost none of these cases is tracked back to its cause. Dr. Michael Osterholm, Minnesota's chief epidemiologist, comments that, "If you get an outbreak of 500 people in a state, but no more than a few in any one household, you'll never pick it up." The Food and Drug Administration currently samples less than one percent of the shipments of 30 billion tons of food imported annually into the United States. Twelve states have no system for reporting foodborne disease, largely because of budget restrictions. It is evident that infectious organisms could be intentionally introduced into the United States with little likelihood of detection prior to the food being eaten. Domestic surveillance for foodborne illnesses is essential not only for tracking natural outbreaks, but for detecting possible intentional poisonings.

Unexplained Deaths due to Possible Infectious Causes

The CDC is establishing an emerging infections program (EIP) network to conduct special population-based surveillance projects. Four EIP sites, in California, Connecticut, Minnesota, and Oregon (covering a population of 7.7 million) are conducting surveys of unexplained deaths and critical illnesses due to possibly infectious causes. The study considers only individuals between the ages of one and forty-nine who are hospitalized with a critical illness due to a possibly infectious cause, with no etiology (disease organism responsible) identified on initial testing. The results of this study are startling. In 1992, 744 unexplained deaths due to possible infectious causes (UDPIC) were identified among previously healthy people in the four sites; these deaths accounted for 14 percent of all 5,304 deaths among persons one to forty-nine years of age in hospitals and emergency rooms. From 1995 to 1997 laboratory specimens and clinical and epidemiological data were collected for UDPIC cases in the four regions and examined by CDC. Yet 77 percent of these cases remained undiagnosed. Some of these may represent fatal infections by altogether new pathogens.

There is evidently a substantial background level of undiagnosed infectious disease in the United States that could be capable of masking sufficiently subtle and dispersed terrorism. From the point of view of improving surveillance for biological terrorism, it is important to recognize that we cannot *currently* recognize what is causing the deaths of many Americans from infectious diseases. Improving surveillance for bioterrorism must begin with the capability to diagnose what is already taking place. Effective surveillance for biological terrorism requires improved surveillance for infectious disease.

Preparing for Biological Terrorism

Preparing for *biological terrorism* has more in common with confronting emerging diseases than with preparing for *chemical* or *nuclear attacks*. Biological terrorism will bypass the quick-response teams that would be critical to coping with attacks using chemical or *radioactive materials*. Unless a biological attack is announced, or discovered while still underway, it would not become clear that something was wrong until victims began to show up in doctors' offices and hospital emergency rooms. The disease agents likely to be used as terrorist weapons may incubate for hours, days, or even weeks before their victims feel any symptoms. More than likely, the earliest symptoms will mimic those of a bad cold or flu. Sufficiently subtle or distributed terrorist attacks may, at least initially, be indistinguishable from naturally occurring infections or outbreaks. This emphasizes the importance of training physicians and other health care workers to determine rapidly, on the basis of the earliest cases, that an unusual infection is involved. Such training is currently still largely missing.

Because of the *incubation delays*, no nation can protect itself by simply screening travelers at its borders. Nor can a country such as the United States hope to inspect more than a small fraction of the food it imports daily. As agricultural markets become increasingly global, the potential vulnerability of nations to food-borne natural or intentional disease will continue to increase.

Protection against both emerging diseases and biological terrorism must instead rely on *disease surveillance*. The synergy between the responses needed to meet these two threats is a powerful one and the United States and global organizations should take full advantage of it. Improving public health surveillance for biological terrorism must have both strong domestic and international components. It also requires better coordination between public health, law enforcement, and intelligence agencies.

How do we measure success in these endeavors? The success of a strategy of prevention is invevitably difficult to prove unambiguously. The more successful a prevention strategy proves to be, the more a metric for success will need to measure surveillance capabilities rather than response. The absence of an undesired, perhaps catastrophic, outcome can never be proven to be due to any particular level of preparedness. Given the importance of prevention, policymakers must become comfortable with this less direct metric of capabilities. Yet it is, after all, a familiar one in the national security realm—as familiar as the *Cold War* strategy of deterrence.

Domestic Measures

Because incubation delay periods for many diseases are longer than international flight travel times, we cannot hope to stop all diseases at the borders of the United States. Nevertheless, screening and quarantine efforts at ports of entry and inspection of food imports provide an important component of public health surveillance. In 1995, the Committee on International Science, Engineering, and Technology

(CISET) of the Clinton Administration's National Science and Technology Council (NSTC) called for the strengthening of screening and quarantine efforts at ports of entry into the United States. With respect to food safety, the Administration issued the *National Food Safety Initiative*, which includes improved coverage for imported foods (as well as for domestic produce, seafood, and livestock).

Recognizing the importance of acting abroad to ensure domestic protection, the Initiative calls for the Food Safety and Inspection Service (FSIS, within the U.S. Department of Agriculture) to provide technical assistance to countries whose products are implicated in food-borne illnesses. These initiatives should improve surveillance for both natural and artificial outbreaks. Further improving domestic surveillance requires improving sensitivity and connectivity along the chain from physicians to national health authorities. We will consider each link in that chain in turn. First note, however, that it may be unrealistic to expect domestic public health agencies to find substantial resources to improve surveillance for biological terrorism within their existing budgets. For example, the formal mission of the CDC is "To promote health and quality of life by preventing and controlling disease, injury, and disability"; its mandate is to provide the greatest good for public health.

It is inevitably difficult to draw resources from programs that are protecting the lives of Americans from day-to-day life-threatening illnesses and redirect them to surveillance for future attacks that may or may not ever take place. Expanding public health surveillance explicitly to include surveillance for biological terrorism will require new resources. An announced biological attack, or one discovered while under way, will require first responders who are appropriately trained. Other biological attacks must first be recognized by pathologists, physicians, and other health-care personnel in family practices, clinics, and hospitals. It is important, therefore, that these medical professionals have some knowledge of the clinical presentations of likely biological terror agents. There is a broad parallel with actions recommended in the Clinton Administration's CISET report for addressing emerging diseases. That document called for expanded formal training and outreach for health-care providers.

The National Institutes of Health (NIH) and CDC have responded by writing to medical and microbiology associations and other professional organizations urging them to focus training and certification programs on emerging diseases, and continue to sponsor meetings on related training needs. Similar actions on the part of NIH and CDC to raise physicians' awareness of biological agents should be undertaken. A first step is the article "Clinical recognition and management of patients exposed to biological warfare agents" in the August 6, 1997, issue of the *Journal of the American Medical Association*. However, the best way to ensure that busy physicians improve their expertise in this area is to require relevant knowledge in medical school curricula and certification examinations, and to offer appropriate training.

In FY97, Congress appropriated $52.6 million to the Department of Defense (DoD) to implement various domestic preparedness programs. These funds were used in a variety of ways, including the development of a Chemical-Biological Rapid Response Team (CBRRT) and to procure additional equipment for the U.S. Marine Corps Chemical Biological Incident Response Force (CBIRF). The DoD also began to train trainers in 120 U.S. cities to prepare for and respond to emergencies involving weapons of mass destruction. By the end of 1997, twenty-seven cities had received visits. The DoD expects to discontinue this training after FY99.

Lead agency responsibility for this training should therefore be transferred to the *Public Health Service* (PHS). The Federal budget currently devotes some $7 billion annually to unclassified terrorism-related programs. It is critical that within this vast budget, sufficient and ongoing resources be found to train the local physicians and other first responders to any likely biological attack.

In June 1998, President Clinton requested an additional $294 million from Congress to deter and respond to terrorist incidents involving biological and chemical weapons. This request included continued funding for local training programs. The first step in improving sensitivity for incidents of biological terrorism is for the federal government to make this a long-term, sustained commitment to training for the nation's physicians, pathologists, and other first responders.

Next, regional centers of excellence, building directly on the best state public health laboratories, should be established with the capability for rapid diagnoses of clinical samples from within their geographic areas. These regional centers must have the trained personnel and diagnostic tools necessary to accomplish this mission, and connections to both local and national institutions must be assured.

To improve the ability to identify rapidly and accurately the early stages of a possible bioterrorist attack, some six to ten regional sites around the United States should be designated for substantial improvement in both epidemiological and rapid diagnostic capabilities.

This capability for high-volume rapid diagnostics differs from the traditional expertise of national reference laboratories. These new regional centers of excellence should build directly on the best of the state public health laboratories in order to minimize additional expense. The President's 1998 request to Congress also asks for an additional $43 million to improve the ability of public health centers to recognize and share information on outbreaks of suspicious diseases. This important request, which if implemented would improve both sensitivity and connectivity, should be fully funded.

At present, too little attention is being given to developing rapid diagnostics appropriate to this sort of laboratory setting. For example, state or regional laboratories will need diagnostics for biological agents that are capable of thousands of sequential assays. Such diagnostics would not have to be hand-held, and would not necessarily need to employ cutting-edge technologies. But they would need to be robust and reliable.

Similar equipment may also be useful for the Army and Naval Research Facilities overseas; at times of major outbreaks of infectious diseases, these regional reference labs may be swamped by samples requiring examination. (The Department of Defense operates infectious disease laboratories in six countries overseas. These labs conduct epidemiologic investigations, diagnose diseases, and recommend control measures. They conduct research on diseases of mutual interest to both the host country and the United States.) Whereas high-volume diagnostics might remain unstressed for long periods in the United States, at reference laboratories overseas they would more likely be challenged by use in real outbreaks; this could provide a valuable opportunity to refine these tools under real conditions. The relevant federal agencies should ensure that at least one agency is working to meet needs for robust, rapid, high-volume laboratory diagnostics.

A related requirement, also with resource implications, is to maintain a cadre of individuals in the United States with expertise in the diseases likely to be employed by terrorists. For example, at present CDC has the only laboratory in the world that serves as a reference laboratory for plague. There is only one full-time employee at that laboratory with experience and training in plague epidemiology and treatment.

The agencies that would be called upon to perform these tasks in the event of outbreaks of biological warfare agents should complete an inventory of critical personnel needs. It is unlikely that agencies will contribute additional positions to individuals with expertise in diseases rarely encountered in the United States. (In 1996, five cases of plague were reported in the United States, of which two were fatal; both decedents died before plague was diagnosed.) Yet individuals with training appropriate to most biological warfare agents are important for responding to outbreaks abroad, since most biological agents are also naturally occurring diseases in one or another region overseas. Funding and positions should be provided for sufficient individuals to maintain national expertise in those diseases likely to be used for biological terrorism.

International Networks

Surveillance for disease outbreaks overseas must also be improved. The surest way to alleviate human suffering, as well as to prevent disease from reaching America's shores, is to detect and stop outbreaks quickly while they are still abroad. The capacities that are needed—trained health care workers and epidemiologists, regional laboratories with reliable diagnostic equipment, good communications, and the ability to send in teams of experts—will help spot both emerging diseases, as well as any outbreaks resulting from the use, testing, or accidental release of biological agents.

The first step should be to improve the existing international network for the detection of *infectious diseases*. There are currently too many geographic holes in the international disease surveillance system. The *World Health Organization* (WHO), the obvious choice for a multilateral solution, has in the recent past been viewed with skepticism by many experts, due in part to its limited resources. "By the time WHO realized there was an AIDS epidemic it already existed on four continents. That's WHO preparedness and emergency response for you," commented D.A. Henderson, the physician who led WHO's smallpox eradication effort.

But there is cause for growing optimism. In 1995, the World Health Assembly, the legislative body of the WHO, adopted a resolution calling on WHO to lead, strengthen and coordinate international efforts to respond to emerging infectious diseases. As a result, the Division of Emerging and other Communicable Diseases Surveillance and Control (EMC) was established, with a mission to strengthen national and international capacity in the surveillance and control of communicable diseases.

The WHO/EMC publishes in both print and electronic formats the bilingual English/French *Weekly Epidemiological Record* and the electronic *Disease Outbreak News*. It is also compiling a searchable database of the WHO/EMC collaborating centers worldwide and, jointly with the World Bank and the Joint UN Programme on HIV/AIDS (UNAIDS), connecting the collaborating centers electronically. Simultaneously, the Program to Monitor Emerging Diseases (ProMED), an international non-governmental group of infectious disease experts, has established an electronic reporting system open to unconfirmed reports of disease outbreaks. This system parallels the more strongly filtered WHO *Rumour Outbreak List*.

These steps are reminders of how "*connectivity*" increasingly means access to electronic mail and the World Wide Web, and the extent to which international health security is enhanced when all nations, including developing nations, gain access to these networks. This is an area where U.S. agencies such as the U.S. Agency for International Development (USAID) may be especially well placed to provide technical assistance and support, working together with international agencies and host governments.

There are two broad categories into which improvements in international surveillance for bioterrorism may be divided. The first is ongoing "*background*" surveillance with the intention of recognizing outbreaks as they occur, while the second involves a directed response to a specific outbreak that has been detected. These latter cases remain in the category of surveillance as long as the "*response*" includes an investigation whose goal is to identify the nature, extent, and origin of a disease outbreak. By this definition, the CDC team dispatched to the Ebola outbreak in Kikwit, Zaire, in 1995 was involved in surveillance (in addition to its critical missions of providing medical care and containing the outbreak).

Investigations of recognized outbreaks (especially those deemed suspicious) and the ability to identify the responsible organism or strain are critical, but these capabilities are dependent upon a surveillance system operating in the background that is able to detect outbreaks as they occur. While egregious attacks or accidents in biological warfare programs may be difficult to miss, the ability to verify the *Biological and Toxin Weapons Convention* (BWC) and to deter would-be violators is enhanced by having as sensitive a public health surveillance network as possible, one that will detect outbreaks that are subtle or identify less-than-subtle outbreaks in their earliest stages. Moreover, such a network provides the best opportunity to stop an outbreak before it reaches the United States.

The surest route to such a capability is to improve the international surveillance system for emerging diseases. In 1996, Vice President Gore announced the Clinton Administration's new policy for responding to emerging infectious diseases. Under that policy, President Clinton directed that the U.S. government would "work with other nations and international organizations to establish a global infectious disease surveillance and response system, based on regional hubs and linked by modern communications technologies."

Such new regional networks should be integrated with the five independent monitoring and alert systems of the WHO/EMC. Information from these systems are made freely available on the World Wide Web and in other fora. One of these systems comprises the WHO Collaborating Centers, a network of over two hundred laboratories and institutions around the world. Collaborating Centers carry out specific activities on behalf of WHO and provide information on disease distribution, while providing laboratory diagnoses and training in the host nation. Host governments agree to allow the Centers to report directly to WHO, without first going through the government. However, there are large regions of the world where these Centers are absent or rare, including Eastern Europe and much of Saharan and sub-Saharan Africa, Central America, and Southeast Asia. The new regional networks would help fill these gaps.

The CDC has thoroughly examined how to support the development of international regional networks of closely linked epidemiology and laboratory programs to promote disease surveillance. These plans were outlined in the CDC document, Addressing Emerging Infectious Disease Threats: A Prevention Strategy for the United States. In 1993, ProMED had endorsed a system similar to that proposed by the CDC. Ten medical centers, strategically located in the developing world, would serve as global health sentinels. The centers would be built directly upon the most capable existing facilities, in order to minimize expense, but would need to be given priority for international assistance. The initial costs for such a network could be modest, perhaps $10 to $20 million per year. This would represent a small fraction of the $7 billion in unclassified terrorism-related programs the U.S. government currently spends. If the United States wishes to improve global surveillance for either emerging infectious diseases or incidents of biological terrorism, taking the lead in developing an international surveillance network is perhaps the most important commitment it could make.

A proposal for substantially augmenting the global monitoring system might also provide a useful tool in the BWC negotiations. Under Article X of the BWC, States that are Parties to the Convention "in a position to do so shall also cooperate in contributing individually or together with other States or international organizations to the further development and application of scientific discoveries in the field of *bacteriology* (biology) for prevention of disease, or for other peaceful purposes."

One element of the global monitoring system being strengthened by the Division of Emerging and other Communicable Diseases Surveillance and Control (EMC) provides a model for public health reporting that sidesteps explicit references to biological agents. The International Health Regulations (IHR) are the only international public health legislation that requires mandatory reporting of infectious diseases (*cholera*, *plague*, and *yellow fever*). To transform the IHR into a global alert system, WHO is revising them to broaden their scope to include many diseases for which they currently make no provision. The approach is to require notification of five specific *clinical syndromes* (respiratory, neurological, antimicrobial resistance, diarrhoeal, and hemorrhagic). Reports of syndromes will be followed by reporting of specific diseases once the diagnosis is known, but action can commence even before a laboratory diagnosis is made.

From the point of view of those concerned with incidents of *biological terrorism*, these five syndromes will capture outbreaks due to biological warfare agents as well as natural causes. This in turn could provide an appropriate way for regional networks to *de facto* participate in surveillance relevant to biological agents without having to do so explicitly.

Biological and Toxin Weapons Convention

A verification regime for the BWC is hampered by the easy availability and dual-use nature of the microbiological technology needed to culture disease organisms. In this light, investigations of unusual or suspicious outbreaks of disease may be the best option for improving verification. The United States should continue to work for the right of the global community under the BWC to investigate suspicious outbreaks wherever they occur. Would-be developers of biological weapons should fear that if an accidental release occurs, the world may discover the resulting outbreak and pinpoint its origin. These same investigations may lead to the identification of unusual but natural outbreaks as well.

The Ad Hoc Group of Governmental Experts (also known as Verification Experts or VEREX) created by the Third Review Conference for the BWC in 1991 explored twenty-one different possible verification measures for the BWC, including surveillance of publications and legislation, scheduled declarations of activities, remote and on-site inspections, and others. The VEREX concluded that no combination of measures could be found that would uncover violations with a high degree of confidence. A consideration of the demands placed by attempted verification of the BWC makes it clear why the task is so difficult. For example, sales estimates of fermenters in the range appropriate for illicit pilot-plant production of biological agents numbered 3,600, in 1984 alone; such plants could be attached to most major universities or biological firms. Because of such practical considerations, BWC negotiators have narrowed the likely categories of declarable facilities to biosafety level 4 laboratories, those facilities producing vaccines or biopesticides, and military and biodefense programs. Otherwise, the number of sites requiring verification is just too large.

Moreover, U.S. companies have concerns regarding the protection of industrial secrets which might be compromised by inspections under a BWC regime. In any case, the experience of the United Nations Special Commission (UNSCOM) in Iraq makes it clear that even comprehensive mandatory declarations and intrusive challenge inspections of a range of biocapable facilities is insufficient to guarantee compliance: Iraq developed and maintained a biological weapons capability while under the direct scrutiny of UN inspectors.

Investigations of unusual or suspicious outbreaks of disease may therefore be the best option for improving verification of the BWC. President Clinton endorsed such a measure in his speech to the UN General Assembly in September 1996. In light of the history of accidental releases in biological weapons programs, a right of investigation could provide a deterrent to such programs, and a possibility of detecting violations. The Defense Special Weapons Agency has conducted a disease outbreak exercise that corroborates the utility of on-site epidemiological investigations in making determinations of the nature of unusual disease outbreaks, and it has outlined preliminary criteria for recognizing possible biological weapons events.

Mechanisms for the initiation of formal on-site epidemiological investigations of suspicious disease outbreaks are under discussion in ongoing negotiations for a protocol to the BWC, and the United States should place high priority on these negotiations. Because it is important that health agencies be able to operate overseas in a transparent manner, directed investigations into suspicious outbreaks are probably best left to teams specifically organized under the BWC.

A different kind of deterrent stems from some of the same molecular biological technologies that could facilitate the engineering of improved biological agents. Genetic fingerprinting (or more broadly, biological signatures tracking: the ability to identify, distinguish, and establish relationships between particular strains of organisms through biochemical or molecular biological analyses of those strains, for example via the development of a library of DNA sequences corresponding to different strains of viruses and bacteria) would help assure would-be attackers that even a secret biological release might nevertheless be tracked to its source. Greater transparency, to include the exchange of strains of organisms held in the national laboratories of individual nations, could facilitate this goal. A DNA-sequence database for different strains of organisms, especially for those associated with weapons programs, is very important for investigating either domestic or international outbreaks. Biological signatures tracking and attribution could be a powerful tool for identifying when an outbreak is artificial and who its perpetrator might be. Biological signatures tracking and attribution research and development should receive high priority for continued and additional funding.

Improved Coordination

Coordination between public health and civilian emergency response agencies is improving. Under the National Food Safety Initiative, the four federal agencies charged with responding to outbreaks of food-borne and water-borne illnesses (the Food and Drug Administration [FDA], CDC within Health and Human Services [HHS], the Food Safety and Inspection Service [FSIS] within the U.S. Department of Agriculture [USDA], and the Environmental Protection Agency [EPA]) are establishing the Food-borne Outbreak Response Coordinating Group (FORCG) to develop standardized procedures for the rapid exchange of data and information associated with food-borne illness outbreaks. The HHS, USDA, and EPA will designate the Assistant Secretary of Health, the Under Secretary for Food Safety, and the Assistant Administrator for Water, respectively, as their outbreak coordinators.

However, public health surveillance, both domestic and international, could also be improved through better coordination among public health and law enforcement and *intelligence agencies*. Coordination is inhibited because of the conflicting demands created by the transparency required for public health agencies to operate freely in the United States or abroad, and the requirements of law enforcement or intelligence gathering. Consider, for example, an institution such as a hospital or university that experiences a disease outbreak. Personnel and administrators may talk freely to scientists pursuing a public health mission, but may be much less forthcoming if those investigators are perceived as surrogates for law enforcement agencies which could pursue possible prosecutions. Internationally, the situation is even more delicate. After the plague outbreak in Surat, for example, the Indian newsweekly *The Week* explicitly accused the United States of being responsible for the outbreak and identified by name four members of the CDC who had arrived in India to study it. The CDC's desire to send epidemiologists was described as suspicious. U.S. agencies conducting epidemiological or other public health activities, be they civilian or military, will be understandably reluctant to risk compromising their ability to detect and respond to diseases overseas by appearing to have ties with intelligence gathering or covert activities.

Nevertheless, the threat of biological terrorism, and potential early ambiguities between natural outbreaks and intentional or accidental releases of biological agents, demand that closer ties between law enforcement, intelligence, and public health be established. For example, public health surveillance could likely benefit from domestic and international intelligence that there was a probable biological threat and consequent concern over the potential use of a particular biological agent. Conversely, law enforcement and intelligence could benefit from being regularly informed on what outbreaks are being seen in public health surveillance (domestic and overseas) and how these events are being resolved.

Federal law enforcement has considerable experience working with local actors (such as local police departments) and success at maintaining the confidentiality of appropriate information passed on in these relationships. The appropriate levels and individuals in the public health surveillance system to receive analogous information need to be determined. Maintenance of confidentiality could be inconsistent with a very broad notification. For this sort of exchange of information to be secure and effective, pre-planning is a requirement.

Role for Scientists and Scientific Societies

Scientists have in the past alerted the public and decision-makers to dilemmas posed by their research, for example in the 1970s during the recombinant DNA controversy. Zilinskas has suggested that, given the difficulties in verifying the BWC, individual scientists and scientific societies must cultivate an ethic in which the illicit development of biological weapons will be discouraged and perhaps revealed to outsiders. He recommends that scientists in nations suspected of sponsoring biological weapons research be especially encouraged to attend scientific meetings and provided with electronic communications access to their international colleagues. Similarly, science students from these nations should be invited to international fora where scientific ethics are discussed. National and international scientific societies such as the American Society for Microbiology and the International Council of Scientific Unions are natural sponsors for such activities. Moreover, scientific and technical workers who once worked in biological weapons programs, for example in the former Soviet Union, should be provided with challenging work in their home nations so as to deter them from marketing their biological weapons skills abroad. International programs established by the European Union, Japan, the United States, and private individuals, such as the International Science and Technology Center, should be supported with these objectives in mind.

National Security and Public Health

Terrorist attacks using *biological weapons* have been carried out or attempted at virtually every scale, from individual assassinations to indiscriminate attacks. While *apocalyptic* urban attacks have not succeeded, they have been attempted by at least one terrorist group. Prudent national security policy requires the United States to prepare itself for such attacks.

Because diseases have long incubation times when compared to modern national and international travel times, preparing for *biological terrorism* is necessarily different from preparing for attacks using other weapons of mass destruction. Preparing for bioterrorism requires improving the sensitivity and "*connectivity*" of public health surveillance systems within the United States and overseas. Domestically, physicians and other health care workers must be given the training needed to recognize or at least suspect unusual diseases, and the ability to check these suspicions quickly at the state or regional level must be available. Internationally, the United States should work with foreign governmental, multilateral, and non-governmental organizations to improve global surveillance for suspicious outbreaks. These same systems will also help protect American citizens, and people throughout the world, from emerging diseases.

The *incubation delay* periods of many diseases as well as the growing amount of food imported into the United States demonstrate the insufficiency of protecting the security of U.S. citizens through monitoring for human-borne illnesses at ports of entry or inspection of food imports. Both forms of monitoring are important and should be improved, but some diseases will inevitably elude this screening. There is no alternative to a defense in depth, with improved surveillance at all levels, from the local to the international. The United States has a strong stake in determining the nature and origin of outbreaks overseas. Threatening outbreaks are best recognized and stopped while still abroad.

An important component of such international detection and response will continue to be the dispatching of epidemiological and medical teams overseas. However, international surveillance must primarily rely upon multilateral cooperation, in the form of both sentinel laboratories and formal and informal electronic and voice networks. It is therefore directly in the interest of the United States to strengthen these.

Preparations for a biological attack via improved public health surveillance, both domestic and international, will simultaneously protect U.S. citizens against emerging infections and other naturally occurring outbreaks. Even if a major biological terrorist attack never occurs, the investment in public health will, on a daily basis, work to improve the health of all Americans.

It is sobering that many of these conclusions are reiterations of lessons learned decades ago. In 1950, soon after the onset of the Korean War, it was recognized that a biological terrorist attack within the United States was possible, and that little could be done to stop such an attack. However, the disease could be contained and quickly treated if early detection were achieved. To this end, the CDC's *Epidemic Intelligence Service* (EIS) was formed; medical officers were trained in field epidemiology and assigned to the CDC, state health departments, and universities. In the absence of a terrorist attack, members of the EIS could maintain their expertise and improve American public health by analyzing natural disease outbreaks. These baseline assessments would in any case be important for the recognition of possible attacks, as they would be needed to determine whether an outbreak exceeded normal background levels—and was therefore potentially suspicious.

An appropriate national security response to the threat of biological terrorism is interwoven with the response that is needed to combat the environmental threat of emerging diseases: improved public health surveillance. In this context, responsibility for national security extends throughout society, from primary care physicians at local hospitals and clinics, to state and national health laboratories and officials, and to public health surveillance networks overseas. These responsibilities need to be matched with appropriate resources. Public health and national security merge in the realm of emerging diseases and biological terrorism.

8

VIRULENCE DETECTION

With the end of the Cold War, the threat of nuclear holocaust faded but another threat emerged—attack by terrorists or even nations using biological agents such as bacteria, viruses, biological toxins, and genetically altered organisms. The former Soviet Union once had a formidable biological weapons program. Now, several countries and extremist groups are believed to possess or to be developing biological weapons that could threaten urban populations, destroy livestock, and wipe out crops.

Even terrorists with limited skills and resources could make biological weapons without much difficulty, says Tony Carrano, Lawrence Livermore's associate director for Biology and Biotechnology Research. "It's not complex, it's not expensive, and you don't need a large facility." For these reasons, biological weapons have been dubbed the poor man's atomic bomb. Contributing to the ease of making and concealing biological weapons is the dual-use nature of materials to produce such weapons, because they are found in many legitimate medical research and agricultural activities as well. CIA Director George Tenet touched on this topic in Congressional testimony in February when he noted the overlap between manufacturing vaccines and producing biological weapons.

The agents used in biological weapons are difficult to detect and to identify quickly and reliably. Yet, early detection and identification are crucial for minimizing their potentially catastrophic human and economic cost. Lawrence Livermore scientists are participating in the Department of Energy's program to improve response capability to biological (as well as chemical) attacks on the civilian population.

A major part of DOE's program is developing better equipment, both fixed and portable, to detect biological agents. However, any detection system is dependent on knowing the signatures of organisms likely to be used in biological weapons. These signatures are telltale bits of DNA unique to pathogens (*disease-causing microbes*). "Without proper signatures, medical authorities could lose hours or days trying to determine the cause of an outbreak, or they could be treating victims with ineffective antibiotics," says Lawrence Livermore's Bert Weinstein, deputy associate director of Biology and Biotechnology Research. Because of the importance of biological signatures, DOE has launched a biological foundations program as a key thrust of its effort to improve response to terrorist attacks.

The program involves experts at the Lawrence Livermore, Brookhaven, Los Alamos, and Sandia national laboratories, as well as colleges and universities. Researchers from the four national laboratories get together at least quarterly to share information and yearly for a formal review of their work. Weinstein reports that important progress has been made since the program began in early 1997, and new signature sets are being transferred to the Centers for Disease Control and Prevention and the DOE. Over the next several years, DOE scientific teams expect to produce species-level signatures for all of the most likely biological warfare pathogens. The teams also expect to have an initial set of species-level signatures for likely agricultural pathogens, because an attack on a nation's food supply could be just as disruptive as an attack on the civilian population.

Several Levels of Signatures

The teams also aim to develop strain-level signatures for the top suspected agents. Strains are a subset of a species, and their DNA may differ by about 0.1 percent within the species. A species, in turn, is a member of a larger related group (genus), and its DNA may differ by a percent or so from that of other members of the genus. Characterizing pathogens at the strain level requires significantly more work than recognizing a species. But strain-level signatures are essential for determining the native origin of a pathogen associated with an outbreak; such information could help law enforcement identify the group or groups behind the attack.

The biological foundations work aims to provide validated signatures useful to public health and law enforcement agencies as well as classified signatures for the national security community. In developing these signatures, biological foundation researchers are also shedding light on poorly understood aspects of biology, microbiology, and genetics, such as *immunology*, *evolution*, and *virulence*. Increased knowledge in these fields holds the promise of better medical treatments, including new kinds of vaccines.

The biological foundations work is one element in DOE's Chemical and Biological Nonproliferation Program. Livermore's component of this work is managed by its Nonproliferation, Arms Control, and International Security Directorate. Other components of the overall program include detection, modeling and prediction, decontamination, and technology demonstration projects. Livermore researchers were among the first to recognize, in the early 1990s, the tremendous potential of detectors based on DNA signatures. "We knew that a lot of work was necessary to develop the signatures the new detectors would need," says Weinstein. In particular, the researchers recognized several *pitfalls*. For example, if signatures are overly specific, they do not identify all strains of the pathogen and so can give a false-negative reading. On the other hand, if signatures are based on genes that are widely shared among many different bacteria, they can give a false-positive reading. As a result, signatures must be able, for example, to separate a nonpathogenic vaccine strain from an infectious one.

Several Levels of Identification

To enhance their detection development effort, researchers are exploring advanced methods that distinguish slight differences in DNA. They are using the multidisciplinary approach that characterizes Livermore research programs. In this case, DNA signature development involves a team of microbiologists, molecular biologists, biochemists, geneticists, and computer experts. In addition, the Livermore work benefits from collaborations with experts worldwide, extensive experience with DNA sequencing, and affiliation with DOE's Joint Genome Institute. Much of the work is focused on screening the two to five million bases that comprise a typical microbial genome to design unique DNA markers that will identify the microbe. The markers, called *primer pairs*, typically contain about 30 base segments and bracket specific regions of DNA that are a few hundred bases long. The bracketed regions are replicated many thousands of times with a detector that uses *polymerase chain reaction* (PCR) technology. Then they are processed to unambiguously identify and characterize the organism of interest.

Weinstein notes that different signatures will be needed for different levels of resolution. For example, authorities trying to characterize an unknown material or respond to a suspected act of bioterrorism will begin with fairly simple signatures that flag potentially harmful pathogens within a few minutes. Typically, such a *signature* would encompass one or two primer pairs and be sufficient for identification at the genus level (*Yersinia* or *Bacillus*, for example) or below.

A *signature* in the next level of resolution is needed for unambiguously identifying a pathogen at the species level (*Yersinia pestis*, for example). This signature involves about 10 primer pairs. Currently, it takes several days to obtain conclusive data for a species-level signature. The goal is to reduce that time to less than 30 minutes.

The *third signature level* is used in pathogen characterization, identifying any features that could affect medical response (for example, harmless vaccine materials versus highly virulent or antibiotic resistance pathogens). This signature level involves some 20 to 30 primer pairs. Together, the primer pairs offer a certainty of correct identification. Currently, providing such a high level of confidence requires several days; the goal again is to reduce the time to less than 30 minutes.

The *final signature level*, intended primarily for law enforcement use, will permit detailed identification of a specific strain of a pathogen (for example, *Yersinia pestis* KIM) and correlate that strain with other forensic evidence. Such data will help to identify and prosecute attackers. The present typical time lag for results is currently a few weeks, and the goal is to reduce that to a few days.

Biological foundations program scientists have worked with DOE and other agencies to assemble a list of natural pathogens most likely to be used in a domestic attack. The list includes bacteria, viruses, and other classes of threats, such as agricultural pathogens. Two extremely virulent pathogens head the list: *B. anthracis* and *Y. pestis*, which cause anthrax and plague in humans, respectively. *Bacillus anthracis* has few detectable differences among its strains, whereas *Y. pestis* strains can vary considerably in genetic makeup. Unraveling the significant differences between the two organisms will give national laboratory researchers experience vital for facing the challenges of the next few years, as they develop signatures for a wide spectrum of microbes.

Livermore Focuses on Plague

Research has been divided and is carefully coordinated among laboratories to avoid duplication. Livermore researchers are focusing on *Y. pestis*, *Francisella tularensis* (a bacterium causing a plague-like illness in humans), and several other microbes that threaten human and animal health. They are working in collaboration with the U.S. Army Medical Research Institute of Infectious Diseases, the Centers for Disease Control and Prevention, the California Department of Health Services, Louisiana State University, Michigan State University, and research centers in France, China, and Russia. "We want to be prepared for the most likely pathogens from throughout the world," says Weinstein.

Eleven species and many thousands of strains belong to the *Yersinia* genus. The most notorious species, *Y. pestis*, causes bubonic plague and is usually fatal unless treated quickly with antibiotics. The disease is transmitted by rodents and their fleas to humans and other animals. Although rare in the U.S., cases are still reported in the Southwest.

Livermore researcher Emilio Garcia notes that the seemingly subtle DNA differences among many *Yersinia* species mask important differences. One species causes gastroenteritis, another is often fatal, and a third is virtually harmless; yet all have very similar genetic makeup. Garcia's team is using a technique called *insertion-sequence-based fingerprinting* to understand these slight genetic differences. Insertion sequences are mobile sections of DNA that replicate on their own. Analyzing for their presence will not only help refine signatures for *Y. pestis* but also shed light on how microorganisms evolve into strains that produce *lethal toxins*. This understanding, in turn, should give ammunition to researchers seeking an *antidote* or *vaccine*.

Garcia's team is collaborating with other world-renowned research centers to better understand the genetic differences among species and strains. A collaboration with France's Pasteur Institute is comparing the genetic complement of *Y. pestis* with another member of the *Yersinia* group (*pseudotuberculosis*) that causes an intestinal disease. "They are closely related, and yet they cause such different diseases," Garcia says.

Better and Faster, with More Uses

Livermore scientists are using a number of methods that allow more rapid identification and characterization of unique segments of DNA. Each method has advantages and drawbacks, with some more applicable to one organism than another. Weinstein expects that within two years, the Livermore team will have settled on a handful of techniques as the workhorses of *signature generation*.

In addition to the insertion sequence method, another promising technique is called *suppressive subtractive hybridization*. The method takes an organism and its near neighbour, hybridizes the DNA from both, and determines the fragments not in common as the basis of a signature. A team headed by Lawrence Livermore biomedical scientist Gary Andersen is working with colleagues at Moscow State University in Russia to advance the technique; one goal is to simultaneously analyze 96 strains of DNA.

Andersen's team has used suppressive subtractive hybridization to distinguish the DNA of *Y. pestis* from that of *Y. pseudotuberculosis*. The team has also used the technique to aid California's poultry industry by providing a handy way to detect *Salmonella enteritidis*. This bacterium can cause illness if eggs are eaten raw or undercooked. *Subtractive hybridization* results have been so successful that the signature can now be used to distinguish between subtypes of *salmonella bacterium*. In addition to the DNA-based pathogen detection methods, researchers are developing detection capabilities using antibodies that can tag a pathogen by attaching to a molecular-level physical feature of the organism.

Antibody assays are likely to play an important role in pathogen detection because they are generally fast and easy to use (commercial home-use medical tests use this form of assay). Biological foundation researchers are working to improve these detection methods as well. For example, a collaboration with the Saratov Anti-Plague Institute in Russia is studying a *bacteriophage* (bacteria-killing virus) that only attacks *Y. pestis* and none of its cousins. Researchers recently discovered that the virus produces a unique protein component to attach to the bacterium cell wall at a certain site and gain entry. Garcia says that recognizing the distinct site could form the basis of a foolproof antibody signature. "If it's possible to achieve it with *Y. pestis*, we may be able to do it with other pathogens," he adds.

SENSING VIRULENCE

As more information about pathogens and their disease mechanisms becomes available and as genetic engineering tools to *transplant genes* become cheaper and simpler to use, the threat of *genetically engineered pathogens* increases. *Biodetectors* must be able to sense the virulence signatures of genetically engineered pathogens, or they will be blind to an entire class of threats. "Our ultimate objective is to identify several specific virulence factors that might be used in engineered biological warfare organisms so that we can detect these engineered organisms and break their virulence pathway," says Weinstein. One key factor useful for detecting engineered organisms is an antibiotic resistance gene. When transplanted into an infectious microbe, the gene could greatly increase the effectiveness of a biological attack and complicate medical response.

Some antibiotic resistance genes are widely shared among bacteria and are easily transferred with elementary molecular biology methods. In fact, a standard biotechnology research technique is introducing antibiotic resistance genes into bacteria as an indicator of successful cloning. "We need to be able to rapidly recognize such genes so that the medical response is appropriate," says Weinstein. Another telltale indication of genetic tampering is the presence of virulence genes in a microbe that should not contain them. *Virulence genes* are often involved in producing toxins or molecules that cause harm or that simply evade a host's defense. "If a series of genes is made available to perform their functions at the right time, they could cause real damage," says Lawrence Livermore molecular geneticist Paula McCready.

If interfering with the action of one of these genes or its proteins interrupts the virulence pathway, the disease process can be halted. Identifying and characterizing important virulence genes and determining their detailed molecular structure will greatly aid the development of vaccines, drugs, and other medical treatments. As an example, *Y. pestis* disables the immune system in humans by injecting proteins into macrophages, one of the body's key defenders against bacterial attack. Because the protein acts as an immunosuppressant to disable the macrophage, understanding its structure not only would help scientists fashion a drug that physically blocks the protein but also would shed light on autoimmune diseases such as arthritis and asthma. A Lawrence Livermore team led by Rod Balhorn is working to determine the three-dimensional shapes of toxins such as the one produced by *Y. pestis*.

Virulence Genes in Common

Virulence genes spread naturally among pathogens and thus are also found in unrelated microbial species. Therefore, virulence genes alone are not sufficient for species-specific DNA-based detection. "We have to differentiate the virulence genes in natural organisms from engineered organisms," says Garcia. Livermore researchers are using different methods for differentiating virulence genes from among the thousands of genes comprising the genomes of pathogens. One technique looks for genes that "switch on" (start making proteins) at the internal temperatures of mammals. For example, Livermore scientists are studying genes of *Y. pestis* that become much more active at 37°C. It seems a safe bet that many of these genes are associated with the bacterium multiplying within a warm-blooded host.

In 1998, a Lawrence Livermore team made an important contribution to understanding the genetics of *Y. pestis*. They sequenced the three plasmids (bits of DNA located outside the microorganism's circular chromosome) that contain most of the *virulence genes* required for full development of the bubonic plague in animals and humans. *Plasmids* sometimes transfer their genes to neighbouring bacteria in what is called *lateral evolution*. (Antibiotic resistance genes are also located on plasmids.) Garcia, who led the plasmid sequencing team, says that studying virulence genes can shed light on how new strains develop. The *Y. pestis* strain that causes bubonic plague, for example, may have evolved some 20,000 years ago. Such understanding is relevant to HIV, which may not have become infectious for humans until the 20th century.

Working with End Users

McCready notes that there needs to be a strong relationship between development of biological signatures and detection technologies and their end uses. Livermore researchers work with agencies that will be using signatures from Livermore and Los Alamos for both hand-held detectors and field laboratories. "We want to make sure our tools get to the experts and agencies that need them," she says.

McCready is working closely with colleagues at the Bioterrorism Rapid Response and Advanced Technology Laboratory of the federal Centers for Disease Control and Prevention. Livermore is collaborating with the CDC to make diagnostic tools available to regional public health agencies and thus create a national mechanism for responding quickly to *bioterrorism threats*. Currently, many health agencies use detection methods that are not sufficiently sensitive, selective, or fast. For example, one culture test for detecting anthrax takes two days. Major damage and even death may have occurred in that time.

McCready emphasizes that DNA signatures will be thoroughly validated before being released, because their use might lead to evacuations of subways, airports, or sporting events, and such evacuations cannot be undertaken lightly. As part of the validation effort, Livermore scientists are characterizing natural microbial backgrounds to make sure that the signatures are accurate under actual conditions. To that end, researchers are collecting background microbial samples in air, water, and soil, as well as in human blood, urine, and saliva.

McCready points out that *B. anthracis* is related to *B. thruginensis*, a naturally occurring harmless microbe that lives in dirt and can give a false positive reading to anthrax if the signature used is not adequately specific. The characterization effort is being aided by a device called the Gene Chip. Manufactured by Affymetrix Inc. and using technology developed by Livermore, the device simultaneously monitors the expression of thousands of genes. Livermore researchers are looking ahead to a time when their efforts will have helped to equip federal and state agencies with a robust set of biological signatures crucial for America's response to any biological warfare threat. Equally important, the researchers envision a strong mechanism linking biomedical scientists with public health and law enforcement officials to develop new signatures speedily and cost-effectively to stay several steps ahead of terrorists.

9

PLAGUE

Yersinia pestis is the causative agent of *plague*, an enzootic vectorborne disease usually infecting rodents (rats) and fleas. Humans can become infected after being bitten by fleas that have fed on infected rodents. In humans, the disease usually occurs in the form of bubonic plague. In rare cases, the infection spreads to the lungs via the bloodstream and causes secondary pneumonic plague. Person-to-person transmission has been described for pneumonic plague but is rare in primary bubonic plague. Bubonic plague can usually be treated successfully with antibiotics; however, pneumonic plague develops rapidly and carries a high fatality rate despite immediate treatment with antibiotics. Plague is also recognized as a potential agent of bioterrorism. It has been used, or considered for use, as a biologic weapon on several occasions. It is important for the medical community to be familiar with the epidemiology, diagnosis, and symptoms of plague so it can deliver an appropriate and calm response should the unthinkable happen.

In recent years, the fear about terrorist attacks with biological weapons has grown. Plague has been identified by the Centers for *Disease Control and Prevention* (CDC) as a category A organism. This third article in a series of papers addressing issues related to biological warfare and bioterrorism gives a concise overview of the role that plague has played in the past and present as a *biological weapon*. As outlined in the historical review of biological warfare, plague has been one of the most devastating epidemic diseases to mankind, second only to smallpox. Given the presence and availability of plague around the world, the capacity for mass production and aerosol dissemination, the high fatality rate of pneumonic plague, and the potential for rapid secondary spread, the potential use of plague as a biological weapon is of great concern.

Medline and *Ovid* databases were searched to identify the pertinent articles and monographs related to plague, *Yersinia pestis*, and biological warfare/bioterrorism. A review of this bibliography led to subsequent identification of relevant references published prior to 1966. The consensus statement of the Working Group on Civilian Biodefense regarding plague was the basis of the final risk assessment and evaluation of the current threat. The final draft published in May 2000 provides a good basis for the development of strategies to counteract the potential threat posed by bioterrorism and the use of plague in particular. However, the conclusions and recommendations need to be regularly reassessed as new information and research become available.

Historical Background and Epidemiology

When the causative organism of *plague* was discovered in 1894, many of the new scientific concepts were subject to lengthy disputes. Naturally, these historic events are now seen retrospectively in light of concepts that are now considered proven. Because of the complexity of the historic background of the disease, this article can provide only a brief summary of the most important historic events.

The oldest account of *plague* is probably given in the Bible, in the First Book of Samuel. This book recounts that in approximately 1000 BC, the Philistines, who had stolen the Ark of the Covenant from the Israelites, were afflicted with a dreadful disease. This disease, which probably was an epidemic of bubonic plague, had afflicted the people in the city of Ashdod, presently in Israel. Eventually, being overpowered by the pestilence, the Philistines were obliged to return the Ark of the Covenant with "five golden emerods and five golden mice." The word "*emerods*" here may denote buboes, and the word "*mice*" may be translated as rats, both supporting the retrospective diagnosis of bubonic plague.

Another report of possible plague is given by Rufus of Ephesus in the first century AD. He describes a plague epidemic in the countries of Libya, Syria, and Egypt. In his account, additional earlier outbreaks of plague are noted, dating back to 300 BC. However, the original records are now lost. More recent literature raises doubts about the true nature of these epidemics. Generally, it is very difficult and in some cases even impossible to render a clear diagnosis from the descriptions of ancient authors. *Smallpox*, *typhus*, and other infectious diseases could have accounted for some of the symptoms. The final states of some of these diseases are quite similar, making it even more difficult to differentiate them retrospectively based on scarce ancient texts.

The first undoubted report of *bubonic plague* is the "Great Plague of Justinian". The disease originated probably around AD 532 in Egypt and spread through the Middle East and the Mediterranean basin in the following years, reaching Turkey, Constantinople, and Greece in AD 541/542, Italy in AD 543, and the territories of France and Germany in AD 545/546. Procopius of Caesarea gives a detailed account of the outbreak of bubonic plague in Constantinople in his book *De Bello Persico*. The estimated population losses in North Africa, Europe, and central and southern Asia were between 50% and 60%. This great first pandemic was followed by many smaller outbreaks throughout the following two centuries, thus bridging the gap between the first and second great pandemics of plague.

In contrast to the *first pandemic*, the second or great medieval plague pandemic is well described by many authors and documents. This second pandemic, also known as the *Black Death* or *Great Pestilence*, appeared in 1334 in China and then spread westward along the great trade routes in Tauris on the Black Sea and eventually to Constantinople. From India it reached the Crimea in 1347 and was then imported into Venice, Genoa, and Sicily. The disease spread slowly and inevitably from village to village by infected rats and humans, or more quickly from country to country by ships, and eventually killed 20 to 30 million people in Europe: more than one third of the European population.

Despite the high mortality rate of the *Black Death* pandemic, the most devastating effects resulted from smaller, recurrent outbreaks that continued well into the 18th century, although with a lower frequency than in the 14th and 15th centuries. Between the years 1349 and 1665, only a few decades saw no plague epidemic. In most cases the epidemics originated from residual foci. In some other cases complete reintroduction of the disease occurred. In continental Europe, three major plague corridors were identified along which the plague epidemics expanded during the 16th to 18th centuries. The first route linked the Low Countries with the Rhineland; the second ran parallel to the Weser and Elbe rivers linking northwestern Germany to Bohemia. The third important corridor was along the coastal region of the Baltic Sea and the North Sea. This *second pandemic*, which lasted more than 130 years, had major political, economic, cultural, and religious ramifications. While many doctors at that time reacted similarly to the Greek physician and philosopher Galen (AD 129–199), who fled when the disease reached Rome, others upheld the highest ideals of the medical profession and continued serving the sick even at their own risk. In the 16th and 17th centuries, many books and tracts were published on the plague and other "*fevers*." However, few contributed significantly to medical progress. On the other hand, many devices and behavioural guidelines were established for those dedicated to the treatment of plague victims.

The first complete theory of infection was developed by Girolamo Fracastoro (1478–1553) and was published in 1546. He proposed that an infective agent of minute size, which he called "*seminaria contagionis*," caused plague. In his theory, the seminaria caused spoiling and were transmitted by minute particles. Although this theory may appear similar to our modern concept of microorganisms, the two cannot be regarded as the same. Only with the invention of efficient microscopes by the Dutchman Antoni van Leeuwenhoek (1632–1723) were microorganisms finally discovered in 1674 and the old concepts of diseases slowly revised. Some 200 years later with the advent of modern microbiology by Louis Pasteur and Robert Koch, many pathogens were discovered, and with the conception of Koch's postulates the widespread theory of a miasmatic cause of disease was finally replaced by the foundation of a scientific bacteriology. However, it took an additional 20 years before the mystery of plague was lifted.

The *third (and present) pandemic* probably originated in the Chinese province of Yunnan around 1855 and spread to the southern coast of China, causing several smaller outbreaks. When the disease finally reached Canton and Hong Kong in 1894, large epidemics occurred, thus marking the beginning of the third pandemic. Plague spread rapidly throughout the world through all inhabited continents, except Australia. Rats aboard the faster steamships that had replaced slow-moving sailing vessels in merchant fleets carried the disease. Between the years 1894 and 1903, plague had entered 77 ports on 5 continents. Since then, smaller outbreaks have occurred around the world. During the early years of this third pandemic, the ultimate death toll in India and China alone was 12 million. In 1900, *plague* was introduced into North America, and between 1900 and 1924 most plague cases in the USA occurred in port cities along the Pacific and Gulf coasts. The disease spread slowly eastward with sporadic cases now being reported mainly in Arizona, New Mexico, Colorado, and Utah. The remainder of cases in the USA are reported in California; cases have been reported in Texas only rarely.

When the plague pandemic reached Hong Kong in 1894, the Japanese government dispatched a commission including the bacteriologist Shibasaburo Kitasato (1856–1931) to investigate this new epidemic, which at that time was spilling over to Japanese ports. At the same time, Alexandre Yersin (1863–1943) was dispatched by the French colonial minister on a similar mission. Both arrived in Hong Kong in June 1894 and independently began their research, ultimately identifying a new bacterium from tissues obtained from dead rats and humans. It is possible that Kitasato was the first to describe the new organism, only a few days ahead of Yersin. A preliminary note appeared in *The Lancet* on August 25, 1894. On the other hand, Yersin's description and explanations published only a few days later seem to be more accurate, with all striking characteristics of the disease emphasized. The literature has been quite inconsistent in crediting Yersin or Kitasato with the discovery of the plague bacillus. Since its discovery, the microorganism causing plague has undergone several nomenclature changes.

Finally in 1970, it was named *Yersinia pestis*. For completeness of a historical review, I shall mention that in 1898 Paul-Louis Simond discovered that plague was transmitted by fleas. In 1927, Ricardo Jorge found the explanation for the endemic occurrence of sporadic cases and outbreaks of plague. He explained that wild living rodents were the infection reservoir of endemic plague. This type of plague was subsequently termed sylvatic plague and has since been described in areas of Russia, South Africa, and South and North America. The risk of importing plague to nonendemic regions may have increased over the past two decades. The worldwide extent of plague-endemic areas and the global incidence of reported cases have both increased, as have the volume and rapidity of national and international travel. In 1991, 1966 cases of human plague were reported; in 1997, the number was 4058. These numbers are the highest for the last 20 years. The recent increase in the number of cases of human plague together with the reappearance of epidemics in countries such as Malawi, Mozambique, and India in 2002 and 2003 led to its recognition as a reemerging infectious disease.

Microbiology

Y. pestis, the causative organism of plague, is a nonmotile, gram-negative bacillus that shows a bipolar staining pattern with Wright,

Giemsa, or Wayson stains. The organism belongs to the Enterobacteriaceae family, is a lactose nonfermenter, and is urease and indole negative. It grows optimally at 28°C on blood agar or MacConkey agar, typically requiring 48 hours for observable growth. The colonies are initially much smaller than those of other Enterobacteriaceae and can therefore be easily overlooked.

The main characteristic of *Y. pestis* is its homogeneity, considering the wide range of hosts and vectors: there is only one serotype, one phage type, and three biovars. Based on historical data and bacteriological characteristics of the strains isolated from remnant foci of ancient plague, Devignat described the Antiqua, Medievalis, and Orientalis biovars, which caused the first, second, and third pandemics, respectively. Recent genetic evidence has reinforced Devignat's original hypothesis. Lucier and Brubaker explained that the *SpeI* DNA patterns of eight strains of *Y. pestis* were closely related to their respective biovars. Rakin and Heesemann independently confirmed those results.

Other studies concluded that several new ribotypes of the biovar Orientalis had originated within the past century: the original *Y. pestis* strain had spread all over the world and also had undergone chromosomal rearrangements, leading to the local emergence of new ribotypes. Thus, it appears that distinct ribosomal RNA profiles of *Y. pestis* may evolve in short periods of time and in specific geographical areas. Isolation of new ribotypes of biovar Orientalis from Madagascar, Vietnam, and India in recent years may be explained by the modification of the original *Y. pestis* strain that had spread through the entire world during the third pandemic. The questions that remain to be answered are whether these new variants have acquired selective advantages in a new environment and, if so, what the nature of the advantages could be. These challenging questions need to be addressed.

Pathogenesis and Clinical Presentation

The pathogenicity of *Y. pestis* results from its remarkable ability to overcome the defense mechanisms of mammalian hosts and to overwhelm them with massive growth. The rapid multiplication of the organism occurs mainly extracellularly. The entrance of the organism into the mammalian host (temperature 37°C) induces the expression of several virulence factors followed by rapid replication of the organism. *Y. pestis* induces an inflammatory response at the site of its inoculation, which is typically the site of a fleabite. From there, the organisms spread via lymphatics to regional lymph nodes. Experimental studies identified several virulence factors that are essential to the survival of

Y. pestis in the mammalian host. Three plasmids have been identified in *Y. pestis*. One plasmid encodes for the low calcium response (LCR) genes, which are responsible for the expression of 12 proteins that act specifically at 37°C and in the presence of low amounts of calcium. These proteins include a secreted protein (V antigen) and 11 surface and secreted proteins called *Yersinia* outer (membrane) proteins (*Yops*). The *Yops* seem to be especially important for the survival of *Y. pestis*: *Yop H* has specific antiphagocytic activity, *Yop E* is cytotoxic, and *Yop M* binds human thrombin. Two other plasmids encode for a plasminogen activator (Pla), bacteriocin pesticin (Pst), murine toxin (Ymt), and the structural gene for the fraction 1 (F1) protein capsule. The F1 capsule is also expressed at 37°C and has antiphagocytic activity against neutrophils and monocytes.

Although *Y. pestis* can survive in macrophages, it can be killed by neutrophils. Therefore, the F1 antigen is essential for the survival of the organism in the mammalian host. These and many other antigens enable *Y. pestis* to survive in the mammalian (human) host by facilitating the use of host nutrients, causing damage to host cells, and escaping phagocytosis and other host defense mechanisms. The initial local lesion and *inflammation* are accompanied by rapid spread and multiplication of the bacteria. The earliest response to the infection is probably a profuse protein- and mucopolysaccharide-rich effusion. This initial vascular phase of the *inflammatory response* is combined with the direct endothelial toxicity of *yersinial toxins*. In later stages of the infection, necrosis leads to vascular destruction and local hemorrhages. These occur without further bacterial invasion of vascular structures. A prominent neutrophilic infiltrate is present; however, because of F1, *Y. pestis* escapes phagocytosis and destruction. And even though macrophages actively phagocytose bacilli, they are unable to kill them. Instead, the *bacillary toxin* destroys the macrophages and other phagocytic cells of the host defense system. The lesions caused by plague result from the local destruction of tissue and from the systemic effects of *endotoxins*. Some of these toxins cause peripheral vascular collapse and disseminated intravascular coagulation.

The infection with *Y. pestis* in humans occurs in one of three primary clinical forms: *bubonic plague* is characterized by regional lymphadenopathy resulting from cutaneous or mucous membrane exposure, *primary septicemic plague* is an overwhelming plague bacteremia usually following cutaneous exposure, and *primary pneumonic plague* follows the inhalation of aerosolized droplets containing *Y. pestis*.

In most cases of naturally occurring human plague (the classic form of bubonic plague), the victims are bitten by plague-infected fleas; however, contamination of open skin lesions with plague-infected material has also been described. The bacteria then migrate through cutaneous lymphatics to regional lymph nodes, where they are phagocytosed but resist destruction. The result is inflammation and swelling in those affected lymph nodes. After the incubation period of 2 to 6 days, patients typically experience a sudden onset of the illness with severe malaise, headache, shaking chills, and fever. Initially lymphadenopathy may not be striking, but with the progression of the disease, buboes are the dominating and prominent feature of the disease. Buboes measure between 1 and 10 cm, and the overlying skin is often *erythematous*. They are extremely tender, nonfluctuant, and warm and are often associated with surrounding edema. *Lymphangitis*, however, seldom occurs. With appropriate treatment in uncomplicated cases, fever and general clinical symptoms resolve usually within 3 to 5 days. The buboes, however, may remain enlarged and tender for many weeks following an otherwise satisfactory clinical recovery.

Primary septicemic plague, which accounts for 10% to 15% of all cases of plague, is an overwhelming and progressive bacteremia in the absence of a *primary lymphadenopathy*. *Septicemia* may also arise secondary to bubonic plague. Septicemic plague affects all age groups; however, the elderly appear to be at a greater risk of developing septicemia. The presence of rapidly replicating gram-negative bacilli in the bloodstream leads to a self-perpetuating immunological cascade that is typically linked with host response to severe injury and bacterial endotoxin. The host response includes a wide spectrum of symptoms, including disseminated intravascular coagulopathy, multiple organ failure, and adult respiratory distress syndrome. Disseminated intravascular coagulopathy can lead to arteriolar thrombosis, hemorrhage in skin and serosal surfaces, and *gangrenous necrosis* of acral regions. *Plague septicemia*, whether primary or secondary, may also result in metastatic infection of other organs and organ systems. Common complications of septicemia include *plague pneumonia*, *plague meningitis*, *plague endophthalmitis*, *hepatic* and *splenic abscesses*, and *generalized lymphadenopathy*.

Pneumonic plague is the most fulminant form of the disease, resulting in almost 100% mortality. The incubation period is usually 1 to 3 days, with a sudden disease onset characterized by chills, fever, headache, generalized body pains, weakness, and chest discomfort. The initial segmental pneumonitis rapidly progresses to lobar pneumonia and then to bilateral lung involvement. Typical pulmonary complications include localized areas of necrosis and cavitation, pleurisy with prominent effusion, and adult respiratory distress syndrome. As the disease rapidly progresses, the most prominent clinical features are cough, sputum production, increasing chest pain, dyspnea, hypoxia, and hemoptysis. These symptoms may be further complicated by concomitant septicemia. Death usually ensues if specific antibiotic therapy is not begun within 18 to 24 hours of the disease onset, and even then the mortality rate remains extremely high.

Pneumonic plague occurs in two distinct and epidemiologically significant forms. Secondary pneumonic plague results from hematogenous spread of *Y. pestis* to the lungs. Patients typically have symptoms of severe bronchopneumonia, chest pain, dyspnea, cough, and hemoptysis. The invasive infection initiates an inflammatory response resulting in bacterial multiplication in the *pulmonary parenchyma*. When this process spills over into the alveolar spaces, it provides a mechanism by which *Y. pestis* can be expelled during coughing episodes. The spread of *Y. pestis* to close patient contacts via respiratory droplet transmission can initiate an epidemic of primary pneumonic plague.

Primary pneumonic plague, the second form in which the disease may occur, initially presents as an infectious pneumonitis with onset of symptoms often within 24 to 48 hours of exposure. Sudden disease onset and rapid disease progression in an otherwise healthy patient is the typical clinical presentation of primary pneumonic plague. In contrast, a patient with secondary pneumonic plague has usually been ill for several days prior to lung invasion and development of pulmonary symptoms. Many patients succumb to their infection before they develop a well-advanced pneumonia. Patients with primary pneumonic plague have an almost 100% mortality and succumb to the disease after rapid progression of the infection. However, primary pneumonic plague is rare in the USA. A 1997 report by the CDC revealed two recent cases of primary pneumonic plague, contracted after handling cats with pneumonic plague. Both patients had classic symptoms of pneumonic plague in addition to gastrointestinal symptoms (nausea, vomiting, abdominal pain, and diarrhea).

Since diagnosis and treatment were delayed more than 24 hours after onset of symptoms, both patients succumbed to the disease. Pneumonic plague must be considered highly contagious whenever it occurs. Person-to-person transmission appears to be most likely in cold, humid environments coupled with overcrowding. *Y. pestis* is not considered truly airborne; person-to-person transmission requires face-to-face exposure within 2 m of a coughing and ill patient.

When patients present initially, often with vague symptoms of an infection, several differential diagnoses have to be considered. *Bubonic plague* can be confused with streptococcal or staphylococcal lymphadenitis, infectious mononucleosis, cat-scratch fever, lymphatic filariasis, tickborne typhus, tularemia, and other causes of lymphadenopathy. Involvement of intraabdominal lymph nodes may mimic appendicitis, acute cholecystitis, or enterocolitis.

Septicemic plague constitutes a medical emergency, which, unless the clinician has good reason to suspect the specific etiology, usually has a working diagnosis that is nonspecific (e.g., sepsis syndrome, gram-negative sepsis). Fortunately, some of the empiric antibiotic regimens for gram-negative sepsis are effective against *Y. pestis*. The most serious point of confusion in the differential diagnosis of plague sepsis may come from the clinical laboratory. For example, an improperly decolourized gram-stain examination of a blood smear or lymph node aspirate may result in the interpretation of *Y. pestis* bipolarity as a gram-positive diplococcus. In addition, automated bacterial identification devices may not code for *Y. pestis* and may also result in misidentification.

Pneumonic plague can be confused with several other causes of severe and acute community-acquired pneumonia, such as *pneumococcal pneumonia*, *streptococcal pneumonia*, and *Haemophilus influenza*. In addition, confusion with *pulmonary anthrax*, tularemia, *Legionella pneumophila*, Hantavirus pulmonary syndrome, and influenza virus pneumonia can occur. For an early diagnosis of plague, a high index of suspicion is required in naturally occurring cases and especially when biowarfare is suspected.

Plague as a Biological Weapon

Although plague was widespread in ancient and medieval times, several outbreaks occurred following the deliberate use and propagation of this disease. During the second plague pandemic, which swept through Europe, the Near East, and North Africa in the 14th century, plague was deliberately used as a weapon during military conflicts. During the siege of Caffa, a well-fortified Genoese-controlled seaport (now Feodosia, Ukraine), in 1346, the attacking Tartar force experienced an epidemic of plague. The Tartars, however, converted their misfortune into an opportunity by hurling the cadavers of their deceased into the city, thus initiating a plague epidemic in the city. An outbreak of plague followed, forcing a retreat of the Genoese forces. This major incident is described by Gabriel de Mussis, a notary born in Piacenza north of Genoa. This technique was repeated with various "*success rates*" during the next centuries.

Advances in living conditions, public health, and antibiotic treatment made outbreaks less likely in the years after the third pandemic. However, the threat of plague being used as a biological weapon remained. During World War II, the Japanese army, Unit 731, is reported to have experimented on plague and to have dropped plague-infected fleas over populated areas in China and Manchuria. In the years following World War II, several countries, including the USA and the former Soviet Union, among many others, performed research on plague as a potential biological weapon. The former Soviet Union focused on the possibility of releasing plague in aerosolized form, thereby eliminating the dependence on the flea vector. In 1970 the World Health Organization (WHO) published a comprehensive report on the outcome of the possible use of biological weapons over populated areas. It was reported that, in a worst-case scenario—the deliberate release of 50 kg of *Y. pestis* in aerosolized form over a city of 5 million—pneumonic plague could occur in as many as 150,000 persons, 36,000 of whom were expected to die from the disease.

Furthermore, plague bacilli would remain viable in the area for 1 hour and up to a distance of 10 km. The expert panel was concerned that in such a scenario significant numbers of city inhabitants might attempt to escape, hence further spreading the disease. While the USA did not succeed in making quantities of plague bacilli sufficient to use as an effective weapon, Soviet scientists were able to produce large quantities of plague organisms suitable for placing into weapons. There is little published information indicating actions of autonomous groups or individuals seeking to develop *plague* as a *biological weapon*. However, in Ohio in 1995, a microbiologist with doubtful motives was arrested after deceitfully acquiring *Y. pestis* by mail.

After the attacks on the World Trade Center on September 11, 2001, the threat of bioterrorism has reemerged. In October 2001, the aforementioned hypothetical concerns regarding the use of *biological warfare* as *bioterrorism* became reality when a Florida man died of *pulmonary anthrax*. Over the next months, another 10 individuals developed symptoms of *inhalational anthrax*. Four of these individuals also succumbed to the disease. Eight nonfatal cases of cutaneous anthrax also occurred. In retrospect it became clear that a series of letters containing anthrax spores, sent through the US mail system, were responsible for this outbreak. Since then, new antiterrorism legislation has been introduced, and appropriate steps have been taken to educate and prepare the public and the medical community to ensue a reasoned response to future threats.

The epidemiology of plague in bioterrorism would differ substantially from that in naturally occurring infections. The organism would most likely be released as an aerosol. An outbreak of *pneumonic plague* would follow, and patients would present with symptoms initially resembling those of other severe respiratory infections. The size of the outbreak would depend on the quantity of biological agent used for the attack, the characteristics of the strain, and the environmental conditions at the time of the release of the organism. Symptoms would most likely occur within 1 to 6 days following exposure, and most people would die quickly after onset of symptoms. The occurrence of cases in areas not known to have enzootic infections together with no known risk factors for infection and an absence of great numbers of dead rodents all indicate the deliberate dissemination of plague.

Treatment Options and Prevention

Streptomycin was approved by the Food and Drug Administration (FDA) for plague; historically, it has been the preferred treatment. When administered early in the disease, streptomycin has reduced the overall mortality from plague to the 5%-to- 15% range. Because US supplies of *streptomycin* are limited, many experts have suggested *gentamicin* as an alternative form of treatment, although it is not FDA approved for this indication. Its efficacy was equal to or better than that of *streptomycin* in some *in vitro* as well as in vivo studies in mice. In addition, gentamicin is widely available, inexpensive, and can be given as a single daily dose. In a contained casualty setting, *streptomycin* and *gentamicin* are the preferred choices for treatment of adults and children, and *gentamicin* is the preferred choice for pregnant women. However, both drugs have to be administered via intramuscular or intravenous injection.

In a mass casualty setting, oral drugs may be needed. The Working Group on Civilian Biodefense has recommended several oral drugs for the treatment and prophylaxis of plague, acknowledging that many are not FDA approved for that indication. These other antibiotics are *tetracycline*, *doxycycline*, *chloramphenicol*, and *fluoroquinolones*. Within the latter group, preference is given to *ciprofloxacin*, which has been shown to be at least as efficacious as *aminoglycosides* and *tetracyclines*. *Chloramphenicol* has been recommended for the treatment of plague meningitis because of its ability to cross the blood-brain barrier. Beta-lactam antibiotics are not effective in the treatment of plague. Antibiotics that have been shown in animal studies to have poor efficacy against *Y. pestis* include *rifampin*, *aztreonam*, *ceftazidime*, *cefotetan*, and *cefazolin*. These antibiotics should therefore not be used in the treatment of plague. Resistance patterns must be considered when choosing an antibiotic for the treatment of *plague*, and antibiotic susceptibility testing should be performed at a reference laboratory because of the lack of standardized susceptibility procedures for *Y. pestis*.

Consensus recommendations were made for special groups based on the clinical and evidence-based judgments of the working group; again, these recommendations do not necessarily correspond to FDA-approved use, indications, or labeling. In contained and/or mass casualty settings, children should be treated with *streptomycin* or *gentamicin*. *Chloramphenicol* is also considered safe in children aged ≥2 years. In mass casualty settings, children aged ≥8 years may be safely treated with *tetracyclines*. Given the adverse effects of tetracyclines and fluoroquinolones, the working group agreed that in mass casualty settings children can be safely treated with doxycycline. Special recommendations have also been made for pregnant women, in whom aminoglycosides should be avoided. However, in cases of severe illness and/ or in a contained casualty setting, treatment with *gentamicin* is recommended for pregnant women. Balancing the risks of pneumonic plague infection with those associated with doxycycline and *ciprofloxacin* use during pregnancy, the working group recommended that in pregnant women *doxycycline* should be used if gentamicin is not available. No specific recommendation have been made for the treatment of immunocompromised patients due to a lack of studies or animal models of pneumonic plague infection in the immunosuppressed population. At this point, the best recommendation is to proceed with the treatment option given to immunocompetent adults and children.

Postexposure prophylaxis for plague should be administered to individuals with close contact (<2 m) with an infectious case and to those who had potential respiratory exposure. The recommended regimen is *doxycycline* or *ciprofloxacin* given for 7 days on the same schedule as for treatment. The working group recommends doxycycline as the first-choice antibiotic for postexposure prophylaxis. In addition, all persons developing a temperature of 38.5°C or higher or with symptoms of a new-onset cough should promptly begin antibiotic treatment. For infants in this setting, tachypnea would also qualify as an indication for immediate treatment. In a mass casualty setting, special consideration should be given to surveillance of the targeted population in order to identify individuals and communities at risk requiring postexposure prophylaxis. Many of these individuals may not be aware of the outbreak and therefore require special assistance.

Currently, no preexposure prophylaxis or vaccine is available for plague. Until 1999, a formalin-killed whole-cell vaccine was available in the USA for military personnel and researchers; however, it was discontinued after studies found that the vaccine was protective only for bubonic plague and completely lacked protection for pneumonic plague. A similar vaccine was in use in Canada, the United Kingdom, and Australia. With the reemergence of the bioterrorism threat, new efforts have been made to develop a new, possibly genetic-based, vaccine. The recent efforts have focused on the development of immunity to the F1 capsular protein and the V antigen. Research in the pursuit of developing a vaccine that can effectively protect against primary pneumonic plague is ongoing in military research institutions in the USA, Israel, and the United Kingdom. Further information on vaccine development is available from these institutions and will hopefully be available through publication soon.

To date, no evidence exists that *plague bacilli* pose an environmental threat to the population. In fact, *Y. pestis* is very sensitive to sunlight and heat and does not survive long outside the host. Although some reports suggest that *Y. pestis* may survive in the soil and decaying animal carcasses, there is no evidence suggesting an environmental risk to humans in this specific setting. In the WHO risk analysis, it was estimated that a plague aerosol would remain effective and infectious for as long as 1 hour after its release. In the setting of a clandestine attack with plague, the aerosol would therefore long be dissipated before the first patient of pneumonic plague would come to hospital emergency departments.

Modern experience with person-to-person transmission of pneumonic plague is extremely limited. In large plague epidemics in earlier centuries, wearing masks prevented pneumonic plague transmission. Given the available historical evidence, the working group recommends that patients should remain isolated for the first 48 hours of antibiotic therapy and until clinical improvement occurs. Other standard respiratory droplet precautions such as gown, gloves, and eye protection should be implemented as well. In mass casualty settings, individual isolation of patients may become impossible. In this scenario, patients with (pneumonic) plague may be cohorted in isolation while undergoing antibiotic treatment. Should there be a need to transport patients to other facilities, the patients should wear surgical masks. Bodies of patients who have died from plague should be handled with strict routine precautions. However, aerosol-generation procedures such as bone sawing associated with surgery or postmortem examination is not recommended, since those activities are associated with a high risk of disease transmission. If such procedures are ultimately necessary, high-efficiency particulate air-filtered masks and negative-pressure rooms should be used.

In recent years, there is increased concern that a possible *bioterrorism attack* with plague might employ a natural or bioengineered drug-resistant strain. Natural resistance of *Y. pestis* to antibiotics was rare; however, in 1995 a plague isolate from Madagascar contained a multidrug-resistant transferable plasmid. The organism produced TEM-1 β-lactamase, chloramphenicol acetyltransferase, and a *streptomycin-modifying* enzyme. Later that year, a second strain was identified with a plasmid that encoded for the streptomycin-modifying phosphotransferase gene, which resulted in high-level streptomycin resistance. Both organisms were shown to contain plasmids that were easily transferred to other strains of *Y. pestis* as well as to *Escherichia coli*. *Y. pestis* is a member of the *Enterobacteriaceae* family, and as such it will be able to exchange genetic material with multiple genera within this family of organisms.

As there are reports that the *bioweapons* operations of the former Soviet Union engineered multidrug-resistant and *fluoroquinolone-resistant* strains of *Y. pestis*, this new evidence of naturally occurring drug resistance in isolates of *Y. pestis* underlines the importance of continuous reevaluation of guidelines for diagnosis and treatment of plague. More research is needed for effective treatment and vaccine development. In the past centuries, plague has caused social and economic devastation on a scale unmatched by any other infectious agent except for smallpox. Although at the present time the organism is not a major health concern, still approximately 2000 cases annually are reported worldwide. It is evident that plague has not been eradicated and will not be eradicated soon. The WHO recently categorized plague as a reemerging infectious disease. Despite the major advances in the knowledge of the disease, in public health, and in diagnosis and treatment that were made since the discovery of the causative agent *Y. pestis*, the main reasons for the persistence of the disease are found in its epidemiology: plague is essentially a disease of wild rodents that is transmitted by fleas. The control of this wild animal population is inherently difficult, since the burrows are most often located in inaccessible areas.

And even if at some point the infected animal reservoir could be completely destroyed, this would not guarantee the extinction of the disease: *Y. pestis* can survive in animal carcasses and litter for several years, thus being a source of reinfection of other rodents. With the new evidence of the reemergence of plague in Africa and India, the possibility of a fourth pandemic has to be considered. Furthermore, the recent emergence of variant strains and the possibility of resistance to current treatment regimens should lead to continuous research on *Y. pestis* and identification of possible new treatment modalities.

In the *era of bioterrorism*, several other issues have to be addressed. The medical community as well as the public should be educated about the basic infectious disease epidemiology and control measures to increase the possibility of a calm and reasoned response if an outbreak should occur. Furthermore, improved culture methods, biosafety facilities, and methods for susceptibility testing are necessary to allow for a more rapid identification of diseases such as plague. Continuous efforts should be made to seek new treatment modalities. This last concern is of great importance, since concerns about bioengineered organisms have been raised. It is in fact highly feasible to construct extremely virulent organisms resistant to standard antibiotics used for treatment and prophylaxis. A defense plan built on prophylactic antibiotics is highly vulnerable, given the fact that multidrug-resistant plague bacilli have recently occurred naturally. Vaccines have been used in the prevention of diseases for many decades and play a central role in the biodefense against a smallpox attack. It seems logical that our current national biodefense strategy must include the development of vaccines against multidrug-resistant strains of anthrax and plague to effectively protect the population.

A threat that is less likely but must be taken very seriously is the creation of genetic constructs through recombination technology. Such organisms—called *chimeras*—would combine the traits of several pathogens to create a highly virulent, transmissible, and multidrug-resistant organism. Alibek and Handelman described work on chimeras being conducted by Soviet military scientists. These organisms would challenge our ability to respond effectively to a public health threat with bioweapons. Most recent epidemics like HIV and severe acute respiratory syndrome have taught us that we can indeed respond quickly to global public health emergencies and develop diagnostic methods, therapies, and, hopefully, vaccines. However, proper education of the medical community as well as the public remains an essential cornerstone to ensure an effective safeguard for tragedies such as *bioweapons attacks*. Let us hope that we will not have to face such a challenge that is caused by the construction of a deadly pathogenic microorganism developed for the sole purpose of killing humans.

Assessing the Risk

The *Yersinia pestis* bacterium that causes plague is stored in microbe banks around the world. But safeguards would make it difficult to acquire a powerful strain. Plague bacteria are sensitive to sunlight, heat, and disinfectants. It's estimated that the bacteria would remain viable and infectious far up to one hour if released into the air. The Soviets demonstrated that the bacteria could be mass-produced as a potential weapon. But the degree of technical sophistication required is very high. Unless treated quickly with antibiotics, pneumonic plague, the form most likely to be used by bioterrorists, kills almost 100 percent of the time. Overall, the death rate is 57 percent.

The very name—which literally means to *strike* or *hit*—conjures deep-rooted fears that date back centuries. It's understandable why: In the fourteenth century, the *Black Death* was responsible for killing a greater proportion of the world's population than any other disease or war in history. Sadly, we can't consign this deadly disease to the pages of a history book. Hundreds of tons of plague bacteria were produced during the cold War as a potential bioweapon. It is indeed a possible through highly unlikely, bioterror threat.

What is Plague?

There are different kinds of plague, but they are all caused by a bacterium called *Yersinia pestis*. The disease is transmitted to humans by the bite of a flea that has fed on an infected rodent. A single fleabite can inject thousands of tiny organisms into the skin, where they make a beeline for the nearest lymph nodes. What happens next determines the specific form of plague a person contracts.

What is Bubonic Plague?

This is the most common naturally occurring form of the disease. It's also the kind believed responsible for the *Black Death*. Once the plague bacteria reach the lymph nodes, they multiply rapidly, causing the lymph nodes to grow swollen and tender. These swollen lymph nodes are called *buboes*, giving *bubonic plague* its name. Because fleabites are most common on the legs and arms, the lymph nodes in the groin and armpits are most commonly involved.

Other Kinds of Plague

In cases where the bacteria are not contained by the lymph nodes, they enter the bloodstream, where they can affect many different organ systems. This is known as *septicemic plague*. When they travel to the lungs and launch an attack there, pneumonia can develop. This is what is known as *pneumonic plague*. Unlike the bubonic form, which can be treated effectively with antibiotics, pneumonic plague is extremely lethal; about half of patients diagnosed with this form of plague die even with antibiotics and the support of modern intensive medical care. And to make matters worse, pneumonic plague is highly contagious, spread person-to-person when coughing examples the bacteria into the air.

It is the ability of the bacteria to cause severe pneumonia and spread from person to person that underscores the threat we face. It's believed that the most likely *bioterror scenario* would involve releasing plague bacteria into the air over a densely populated area, where it would be inhaled by thousands, who may then spread it to others. In an outbreak of plague that results from such a scenario, many people would develop severe respiratory symptoms and pneumonia over a relatively short period of time. This would present a very different pattern from a naturally occurring outbreak of plague and could initially be mistaken for an outbreak of influenza.

Symptoms of Plague

The symptoms depend on the form of plague a person contracts. Since pneumonic plague is the form most likely to result from a bio-terror attack, we'll deal with it first. Following an aerosol or airborne exposure, the disease progresses extremely quickly, which is one of the main challenges it poses. Like most diseases, the earlier plague is diagnosed and treated, the more likely the infection can be controlled and the potential for an epidemic minimized.

The incubation period—the interval between exposure and the development of symptoms—can range from one to six days, but the first signs and symptoms are usually seen within two to four days from the time the plague bacteria is inhaled. These include *pneumonic plague* after handling infected cats, the diagnosis was not made and treatment didn't start until more than twenty-four hours after the first symptoms appeared. Both patients died.

With *bubonic plague*, it's usually two to eight days before the first symptoms appear: sudden fever; chills, and weakness. Within a few days, the lymph nodes, usually in the groin and armpits, swell and become extremely painful. If the patient doesn't get appropriate care, respiratory failure, shock, and death can follow in two to four days. Rarely, the bite from an infected flea will lead directly to septicemic plague, in which the bacteria enter the bloodstream, severely damaging organs and causing blood vessels to hemorrhage. Sometimes, gangrene may occur in the nose and fingers, and there may be bleeding from the nose and ears.

Without treatment, *septicemic plague* is almost always fatal. In United States over the last fifty years, persons with bubonic or septicemic plague developed pneumonia in about 12 percent of the cases. The last known case of one person spreading plague to another in the United States occurred in Los Angeles in 1924.

Natural Occurrence of Plague

The bacterium that causes plague can be found on every populated continent except Australia. Over the last fifty years, an average of 1700 cases has been reported annually worldwide. Advances in living conditions, such as control of rodent populations, coupled with the availability of antibiotics make naturally occurring outbreaks of plague far less likely today than in the past. But since animals such as ground squirrels, prairie dogs, and occasionally rabbits and cats may be infected with this bacterium, human infection remains a possibility.

Periodic biological surveys of rodents help public officials assess hood that this infection will emerge in the wild. This information is used to provide the public with guidance regarding the likelihood of contracting plague in specific geographical and ecological settings.

In the United States *bubonic plague* remains by far the most common form and occurs primarily in the western states of New Mexico, Arizona, Colorado, and California. Of the 390 cases of plague reported in the United States from 1947 to 1996, the vast majority (84 percent) were the bubonic form and only 2 percent developed into pneumonia. But when pneumonic plague does surface, public concern is heightened. In 1994, the occurrence of a small number of cases of plague in Surat, India, led to panic and fear that an epidemic was about to explode, and approximately half a million people fled the city.

While this fortunately did not develop into a broader outbreak, it highlighted the need for epidemiological and laboratory capacity and expertise to conduct investigations whenever a case of plague occurs. Because of the potential for pneumonic plague to rapidly expand into the epidemic, it is absolutely crucial to quickly determine the source of an outbreak so that appropriate control measures can be implemented.

In addition, the investigation can assess whether the outbreak of plague cases represents a naturally occurring infection or could be the first sign of an intentional release. Information gathered in the course of the investigation is also necessary to keep health care works, the community the media, and local and national decision makers updated as part of the effort to manage the evolving public health crisis.

How Contagious is Plague?

Again, it depends on the form. Pneumonic plague is easily passed from person to person when someone who is infected coughs bacteria into the air. The disease also is considered highly infectious, meaning it is easily contracted from an intentional release of the bacteria into the air. *Bubonic* and *septicemic plague* are spread only through bites from infected fleas, so people who develop this form of the disease are not contagious to others.

Because people who have pneumonic plague are highly contagious, they should be placed in respiratory isolation. Hospital staff should wear respirators—masks that can filter out particles—when they are around patients. Because of the possibility that the disease can be spread to those who come into contact with a pneumonic plague patient, the local and state health departments should be notified immediately.

Reporting a case of plague to local health authorities is required by the World Health Organization's international health regulations so that community monitoring can begin immediately to determine whether the disease is part of a larger outbreak. In addition, an epidemiological investigation is performed in an attempt to find the source and risk factors of the disease so others can be shielded from exposure. Initial laboratory tests, including a determination of the antibiotics that are likely to be most effective, should be confirmed by a referral laboratory in the state or sent to the Centers of Disease Control and Prevention.

Plague as Weapon of Mass Destruction

First, most experts believe this is an unlikely scenario. It would be fairly difficult to obtain the right strain of plague bacteria and even more difficult to process it as a weapon of mass destruction. But we know that thousands of Soviet scientists in over ten research institutes worked with the bacterium that causes plague. Moreover, they produced hundreds of tons of plague bacteria during the Cold War as a potential bioweapon, so we can't delude ourselves into thinking it can't happen.

In a 1970 analysis, the World Health Organization estimated that if fifty kilograms of plague bacteria were released upwind of a city of 5 million, as many as 150,000 people would develop pneumonic plague, 80,000-100,000 would need to be hospitalized, and 36,000 of those would die. This worst-case scenario depends on a number of factors, including the quantity of bacteria released, the dissemination technique, and meteorological conditions. Once released into the air, the bacteria would remain capable of infecting people for up to one hour, but would then be destroyed by the sun's ultraviolet light. But even if the release were not perfectly executed, the effects of the disease and subsequent social disruption would be substantial.

An interactive bioterrorism preparedness exercise was conducted in Denver in 2000 to gauge how ready top government officials and state and local public health agencies are to respond to a bioterror attack. The drill, called TOPOFF (short for "*top officials*"), was based on the hypothetical release of plague bacteria at the Denver Performing Arts Center three days before the beginning of the exercise.

It quickly became apparent that the public health system was not ready to handle such an event. It virtually crippled the city and completely overwhelmed hospitals in the first twenty-four hours. There was confusion over who was in charge, how decisions would be made and communicated, how lifesaving antibiotics and simple face masks would be distributed, how hospitals would be supplied and staffed (medical, administrative, and security personnel), and how mandatory quarantine and isolation could be enforced.

By the end of the exercise, there was civil unrest, rioting, and gridlock. It was not even clear what the human toll would be. Conflicting reports estimated the number of sick persons at 3,700 to more than 4,000, and the number of dead at 950 to more than 2,000.

This was just a simulation, but it opened our eyes to some of the main challenges we face in dealing with the threat of bioterrorism. Identifying where we need to make improvements was extremely valuable. We found out we were under-prepared to deal with such an event. But the drill pointed us in the right direction, especially regarding the need to be able to rapidly increase the number of public health workers available to help control the epidemic with out disrupting their other key duties. Officials at the federal, state, and local levels have already begun addressing these crucial questions. And Sen. Edward M. Kennedy and I joined to write and sponsor (along with more than eighty of our Senate colleagues the Bioterrorism Preparedness Act of 2001, which is designed to improve our health system's ability to respond to bioterrorism, among other goals.

Vaccine against Plague

An old vaccine has been discontinued because of its many side effects. Also, although it helped prevent or ease symptoms of bubonic plague, it was regarded as ineffective against the inhaled form most likely to be used in a bioterror attack.

Work has been under way since the Persian Gulf War on safer, more effective vaccines. That work got a boost last fall when scientists finished decoding the full *DNA*, or *genome*, of the bacterium that causes plague. The sample used to decode the genome came from a Colorado veterinarian who died from the inhaled form of the plague—from an infected cat—in 1992. So it's hoped that vaccine researchers will be able to use this new DNA map to develop a vaccine that protects against pneumonic plague.

Effective Treatment for Plague

The bacterium can be thwarted by a number of antibiotics, but treatment must begin within twenty-four hours after symptoms first appear for the patient to have the best chance of survival. Again, in making the diagnosis and beginning treatment, every moment counts. *Streptomycin* has historically been the first-choice antibiotic against plague. Administered by injection early in the disease, the drug has cut the death rate to between 5 and 114 percent.

But *streptomycin* has been infrequently used in the United States, and large stocks are not available. *Gentamicin*, also administered by injection, is widely available. Physicians have extensive experience using this antibiotic, and it is relatively inexpensive. It has been shown to be at least as effective as streptomycin in treating plague.

Doxycycline and *ciprofloxacin* (Cipro), which can be taken by mouth, also are effective alternatives. They would be the top choices if a bioterror attack resulted in mass casualties, because they can be dispensed in pill form and could be given to those exposed before the infection was advanced. The logistics of trying to administer shots to large numbers of very ill people would be extremely difficult.

Further complicating matters is the fact that patients with *pneumonic plague* would require enormous hospital resources. They would have to be isolated for at least the first forty-eight hours, or until their symptoms improved and they were no longer contagious. However, since this type of pneumonia is often very severe, many would require respiratory-support machinery and other support.

In the event of an outbreak of pneumonic plague, it's a good idea for anyone at risk of exposure to wear an N95 respirator, a mask that must be fitted properly to provide maximum protection. For more information on the respirator, lever, fatigue, and a cough with bloody or watery sputum, a mixture of saliva and mucus. Nausea, vomiting, stomach pain, and diarrhea also are frequent early symptoms.

Because these initial symptoms are similar to those of many respiratory infections, the health care community will need to be alert to the potential threat and be up-to-date about the symptoms patients exhibit and the appropriate management of patients suspected of having pneumonic plague.

Following these early symptoms, the disease progresses rapidly into pneumonia. If the disease is untreated, death can follow with in a day or two. In the days before antibiotics, the time from inhaling the plague bacteria to death was reported to have been just two to six days. In two recent U.S. cases where people contracted.

Plague through the Ages

There are reports of plague like disease in China as early as 224 B.C. In A.D. 541 the first recorded plague epidemic claimed the lives of up to 60 percent of the population of North Africa, Europe, and central and southern Asia.

The second plague epidemic is the one flint came to be known as the *Black Death*, It raged process Europe from 1346 to 1352 and was spread to humans primarily through infected rodents and their fleas. During these six years the Slack Death killed more than One-third of Europe population, accounting for twenty million to thirty million death. The disease moved inexorably from village to village and then was carried by far away as Iceland and Greenland, where the disease may have battened the demise of the Vikings.

The *Black Death* took its name from the gruesome appearance of its victims. Some believe the name derived from victims who turned dark purple in their final hours because of respiratory failure. Others say it may have originated from the gangrene that set in on the noses and fingers of some victims toward the end.

The third worldwide epidemic began in China in 1855, spread to every populated continent, and claimed the lives of more than twelve million people in India and China alone.

Pneumonic plague outbreaks have been rare, but extremely deadly. An outbreak in Manchuria in 1910-11 spread the disease to as many us sixty thousand people. A second large outbreak of pneumonic plague occurred there ten years later. In those days before antibiotics, nearly 100 percent of the cases were fatal.

Are We Ready for the Next Plague?

Microbe: Are We Ready for the Next Plague? by Alan Zelicoff and Michael Bellomo is a comprehensive, yet succinct, account of the threat to public health posed by microbial pathogens. What distinguishes this book from the surfeit of recent books hyping the threat of bioterrorism are its balanced perspective and elucidation of naturally emerging disease threats, such as severe acute respiratory syndrome (SARS) or West Nile virus, as exotic entities requiring a rapid and effective response; Mother Nature is quite the bioterrorist herself. Early recognition that an event has occurred is key to containment of the nascent epidemic.

Zelicoff provides sufficient basic science background to bring the uninitiated up to speed on a variety of exotic and recently introduced microbes in engagingly titled chapters such as "The *Birds that Fell from the Sky*" (West Nile), "*Corona of Death*" (SARS), and "*Something in the Water*" (*Cryptosporidium*). In addition, they describe hantavirus pulmonary syndrome, mad cow disease, and Legionnaires' disease in the context of recent public health emergencies. The authors also explain why both smallpox and anthrax are more than abstract concerns as agents of bioterrorism, on the basis of weaponization history, intrinsic attributes, and realistic scenarios. An account of the 1970 smallpox outbreak, which occurred in Aralsk, Kazakhstan, as a consequence of open air testing of a smallpox weapon by the Soviets is an eye-opener; there should be no doubts about capability and intent after reading this story.

The scenarios are well chosen and informative; they highlight the importance of early recognition that "something has happened" and breaking the disease cycle close to the index case. The unifying theme of the book is the importance of syndrome-based surveillance in achieving this goal. The authors dismiss BIOWATCH (airmonitoring devices to detect and identify microbes in aerosol clouds) as a well-intended but expensive "work in progress," to put it charitably.

BIOSENSE is a national surveillance system that they say has not been implemented in any substantive way.

Healthcare providers recognize syndromes, not microbial diseases. How long did it take to recognize monkeypox in 2003? The hantavirus associated with lethal pulmonary syndrome in New Mexico in 1993 was recognized only when the pattern emerged among previously healthy young adults living in rustic conditions on a Navajo reservation. The authors describe a product they dub Syndromic Reporting Information System (SYRIS) as a "*beta test*" product that has been deployed on a limited, regional basis and promises to provide a near instantaneous map of syndromic reports and to comply with all Centers for Disease Control and Prevention requirements for electronic reporting systems. Like most good ideas, simplicity is central to the SYRIS concept; it is likely to succeed because participating doctors, nurses, and veterinarians (most of the exotic pathogens are zoonoses) can report syndromic occurrences in 15 seconds or less and will be rewarded with instantaneous feedback and tailored reports and alarms. While this section does read a bit like an infomercial, the concept is sound and worthy of serious consideration by public health officials and policy makers. This book is the best of its genre and is recommended for anyone interested in understanding and managing the risks associated with emerging microbial threats.

AIDS Pandemic: Impact on Science and Society

As we enter the third decade of the AIDS pandemic, numerous texts explore the many aspects of AIDS and its consequences. Mayer and Pizer's premise is that AIDS has transformed many of the disciplines that it has touched. For the most part, this wellwritten volume supports their thesis. The authors, all established researchers, tackle many of the major issues, including virology, immunology, vaccines, microbicides, and sexually transmitted diseases, as well as the global impact of HIV/AIDS. Each chapter provides a well-referenced overview of its topic with many references as recent as 2003.

One of the real strengths of this book is a chapter on quantitative science that explores not only the history of HIV clinical trials, but also the design and importance of clinical trials in general. This chapter should be required reading for those considering clinical research in HIV. The chapters on Africa and Asia ably contrast the differences in these areas of highest prevalence. Another strength is the discussion of HIV in correctional facilities and the challenge of caring for this population, including their coexisting conditions and illicit drug use.

Lastly, the discussion of the economics of AIDS is especially welcome in this era of efforts to increase access to drugs worldwide.

Overall, this book fills a valuable niche. A relatively concise text, it reviews many aspects of HIV with a focus on how each topic has evolved over the years. A few tables are small, but overall the diagrams and charts are clear and legible. This book would be of interest to infectious disease fellows, HIV caregivers, and those involved in public health and health policy.

Tick-Borne Diseases of Humans

During the past 2 decades, the scientific landscape of tickborne diseases has changed remarkably. In part because of advances in molecular biology, more than 10 new rickettsial diseases, several ehrlichial diseases, and novel agents of *Borrelia* and *Babesia* genera have been recognized. This renaissance of interest in tickborne infections benefits from advances in molecular phylogenetics and diagnostics, immunology, and informatics that provide tantalizing insights into the complexities of vectorborne infection. Tick-Borne Diseases of Humans is a well-referenced textbook that encompasses these new insights in vector biology and reviews the emerging epidemiology and clinical science of these diseases as they occur across the globe. The editors' goal of providing a "*comprehensive*" resource is admirably fulfilled.

Tracking Down Virulence

Plague is potentially a deadly agent of bioterrorism. Unlike anthrax, which has been so much in the news lately, plague is highly infectious and can be readily passed from one person to another. The bite of a plague-infected flea or the inhalation of just a few cells of plague bacterium can kill. Like smallpox, plague can spread and kill large numbers of people very quickly. Fortunately, it can usually be treated with antibiotics.

History tells us how devastating a plague epidemic can be. In what is known as the Justinian epidemic, from 540 to 590 AD, plague spread from Lower Egypt to Alexandria to Palestine and on to the Middle East and Asia. At its peak, 10,000 deaths occurred every day in Byzantium. Eight hundred years later, in 1347, plague came to Italy from Asia or Africa, probably by ship. By 1351, fully one-third of Europe's population had died from bubonic plague.

This European epidemic is known as the Black Death or the Great Pestilence. In 1894, when Andre Yersin identified the tiny bacterium that causes plague, he named it *pestis* after the Great Pestilence. He tried to name the genus *Pasteurella* after his mentor, Louis Pasteur. But *Yersinia*, after its discoverer, is the name that stuck.

Today, *Yersinia pestis* is one of several infectious diseases and agents of bioterrorism that researchers across the Department of Energy complex are studying as part of the Chemical and Biological National Security Program. This program comes under the purview of DOE's National Nuclear Security Administration (NNSA). At Livermore, the work on *Y. pestis* also receives support from Laboratory Directed Research and Development.

Scientists at Livermore have developed DNA signatures for *Y. pestis* that can be used to quickly detect and identify plague outbreaks. Signatures for nine strains of the disease have been submitted to the Centers for Disease Control and Prevention in Atlanta, Georgia, where they are undergoing a rigorous validation process. Livermore's DNA-based detection method proved its mettle in northern Arizona last June when it was used to identify a plague outbreak in prairie dogs in just four hours. Standard detection processes, which require growing the suspected bacteria in a laboratory, take 36 to 48 hours.

For a plague detector to be truly effective, it must do more than simply indicate the presence of a specific organism known to cause plague, says Pat Fitch, who leads Livermore's Chemical and Biological National Security Program. The detector also must be able to identify the specific traits found in atypical plague-causing organisms. Scientists know of several hundred strains (or isolates) of *Y. pestis*, and they do not all behave in precisely the same way. A few strains are believed to have been genetically modified or engineered to be more deadly. There have also been two clinical cases of naturally occurring antibiotic-resistant plague. Knowing the precise identity of a strain of plague—or of any infectious disease, for that matter—could help physicians treat a patient properly.

Plague research at Livermore currently is focusing on what makes *Y. pestis* so virulent and able to overcome the defenses of a host organism. Fitch is leading the Pathogen Pathway Project, using plague as a prototype for the functional genomics of a larger set of pathogenic agents that could be used in biological terrorism.

Besides building better detectors, work on the *Y. pestis* genome will also lead to a better understanding of pathogenicity and better vaccines and treatments for the disease. The ultimate goal, says Fitch, "is to produce a computer model that simulates the workings of a cell so that we can better manage exposure to pathogens."

Plague as Prototype

The highly contagious *Y. pestis* is an excellent model for studying the interactions of a pathogen and its host. In the case of *Y. pestis*, the host may be a flea, a rodent, or a human. Fleas carry plague bacteria and help transmit the disease. Once an infected flea bites a rodent or human, the bacteria begin to multiply in the new host, and their virulence shifts into high gear. *Y. pestis* circumvents the host's defenses by injecting into host cells a series of virulence factors that inhibit the response of the immune system.

Earlier research has shown that when *Y. pestis* is grown at the body temperature of a flea (26°C), its cells divide, but it does not express (turn on) many of the genes that make it virulent in rodents and humans. When the temperature increases to 37°C (human or rodent body temperature), the bacterium begins to produce the proteins essential to its virulence. This virulence mechanism can be induced in the laboratory, making plague relatively easy to study.

Examination of the *Y. pestis* genome before and after virulence has been induced shows what genes have been turned on. But that information is not enough to show precisely which genes are responsible for various aspects of virulence.

For comparative purposes, a Livermore team led by microbiologist Emilio Garcia collaborated with the Institut Pasteur in France to sequence *Y. pseudotuberculosis*, the parent organism of *Y. pestis*. Although their DNA sequences are about 95 percent identical, *Y. pestis* and *Y. pseudotuberculosis* behave differently. *Y. pseudotuberculosis* lodges in the intestine and causes flulike intestinal distress. *Y. pestis* is also closely related to the mild-mannered *Y. enterocolitica*, an intestinal bug that is itself very much like *Y. pseudotuberculosis*. *Y. enterocolitica* is currently being sequenced by the Sanger Center in Great Britain.

"Bacteria evolve very efficiently and make use of about 80 percent of their DNA," says Fitch. By comparison, humans use only about 30 percent of their DNA. Aiding speedy evolution are the many insertion sequences in a bacterial genome. Insertion sequences are bits of DNA that allow large regions of DNA to replicate themselves and move around the genome, relocating themselves somewhere else. When an insertion sequence lands within a gene, it deactivates that gene. These transfers can also occur across species, and it is not difficult for a bacterium to grab DNA from another bacterium.

Y. pestis evolved from *Y. pseudotuberculosis* within the past 15,000 years, a rapid evolution even for bacteria. "Something happened then to cause *Y. pestis* to learn how to live in a flea," says Garcia. In addition to their normal chromosomal DNA, bacteria may have smaller circles of DNA known as plasmids. Plasmids replicate separately from chromosomal DNA and often house genes that encode enzymes critical to the host cell or organism. For example, when a bacterium has become resistant to antibiotic drugs, it is usually because the bacterium has acquired a new plasmid.

One *Y. pestis* plasmid encodes at least two genes that allow *Y. pestis* to survive in fleas. Another plasmid is home to the gene that activates the disease's invasiveness. Researchers have found that *Salmonella* has a similar plasmid, which one bacterium probably obtained from the other.

"The interesting thing is that if you insert the three *pestis* plasmids into *Y. pseudotuberculosis* or *Y. enterocolitica*, you don't get *pestis*," says Garcia. "So something else is going on. Unfortunately, it's never simple."

Once its virulence genes have been turned on, plague infects its host using what is known as Type III secretion, an injection mechanism more colourfully called "*Yersinia's* deadly kiss." *Salmonella typhi*, enteropathogenic *Escherichia coli*, *Chlamydia psittaci*, various species of *Bordetella*, and other pathogenic bacteria appear to share this syringe-like injection mechanism. This common trait may indicate another area of transferred genomic material.

Before Livermore's research on plague started, many of the genes critical for virulence had been identified but were poorly understood. The same was true for the underlying mechanisms of virulence. There was also little understanding of the gene and protein interactions that take place between the pathogenic bacteria and its host.

The *Pathogen Pathway Project* is using functional genomics tools to identify genes important to virulence and understand the pathways of virulence. The team's hypothesized pathway, from DNA to the host organism, is shown in figure 9.1 at right.

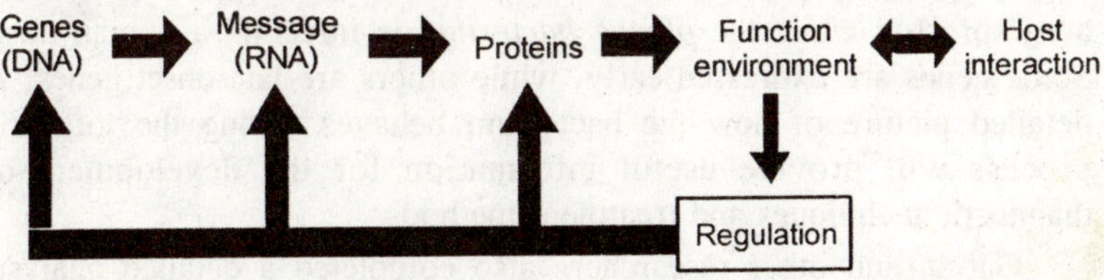

Fig. 9.1. A schematic diagram of information that is hypothesized to describe the pathways of virulence in a pathogen.

Expression and Function

An early task for Livermore bioscientists and computations experts was to develop a relational database of the DNA sequence of *Y. pestis*. In collaboration with the DOE Genome Consortium at Oak Ridge National Laboratory, these data were used to computationally predict where the 4,500 genes in *Y. pestis* are located and which genes might be associated with virulence.

Next, Livermore bioscientist Vladimir Motin and colleagues designed chemical reagents for extracting over 300 genes from *Y. pestis* DNA, including all known virulence-associated genes on the plasmids. In an initial test, they extracted 85 genes associated with virulence and spotted them on a glass microscope slide alongside 11 control spots, making up a 96-spot microarray.

A microarray permits scientists to study the response of thousands of genes or other pieces of DNA quickly and efficiently in a process known as transcript profiling. In the process, each gene receives some kind of stimulus, causing it to turn on and produce messenger RNA (mRNA). In the case of plague, the stimuli are changes in temperature and calcium concentration. The production of mRNA leads, in turn, to the synthesis of unique proteins. The level of mRNA can be measured for each individual gene. The more active or expressed genes there are, the more mRNA will be present.

For the 96-spot microarray, the team developed a protocol to study the response of *Y. pestis* genes under conditions that mimic the infection process: at both flea and human/rodent body temperatures, 26°C and 37°C, and at calcium levels that correspond to those of blood (higher level) and organs (lower level), the latter location being where more virulence genes are expressed.

More recently, they developed a microarray for all 4,500 *Y. pestis* genes. All of the genes are being mapped at six time intervals as temperature rises and calcium concentration drops. The team is thus beginning to establish a timeline for how and when genes change and are expressed while the *plague bacterium* is infecting a human host. Some genes are expressed early, while others are late-onset genes. A detailed picture of how the bacterium behaves during the infection process will provide useful information for the development of diagnostic techniques and treatment methods.

Garcia and other researchers also completed a detailed analysis of three *Y. pestis* plasmids, which allowed them to confirm the location of several known virulence genes and to uncover four novel ones believed to contribute to virulence. Computerized comparisons with other genomic databases indicated the presence of a large number of virulence-related genes that are similar in both closely related bacteria such as *Y. pseudotuberculosis* and distantly related bacteria such as *E. coli*. The team also found numerous gene coding regions whose function they could not determine.

Using a proteomic approach of protein separation techniques and *mass spectrometry* (MS), Livermore researchers led by Sandra McCutchen-Maloney are analyzing complex mixtures of proteins isolated from *Y. pestis*. By comparing samples grown at the two physiological conditions mimicking the flea and the human (at 26°C and 37°C, respectively) and at low calcium concentration to induce virulence, the team is detecting differential protein expression to identify candidate proteins important for *Y. pestis* pathogenicity. Comparisons are also being made between human cells that have and have not been exposed to *Y. pestis* in order to understand the host immune response. Because it is the proteins that are actually responsible for virulence effects, the group is also working to correlate their *proteomic data* with *genomic data* obtained from microarray experiments. To learn more about the individual proteins responsible for virulence, the team is using various biochemical assays to test functional models of the candidate virulence factors. In addition, McCutchen-Maloney's group is looking at host-pathogen interactions by using surface-enhanced laser desorption ionization (SELDI) MS to study various protein–DNA and protein-protein interactions within *Y. pestis* and between *Y. pestis* and the human host. For example, regulatory proteins that bind to genes and control differential expression are under investigation, as are the specific protein–protein interactions of suspected virulence factors. These molecular interactions are key to the genetic feedback that occurs as a pathogen infects its host.

Differences are Key

Before Garcia's team completed its comparative sequencing of *Y. pseudotuberculosis*, microbiologists Gary Andersen, Lyndsay Radnedge, and others examined the differences between *Y. pseudotuberculosis* and *Y. pestis* using a different technique. This process, developed in Russia, is known as suppression subtractive hybridization (SSH). SSH identifies regions of DNA that are present in one species but absent in another.

SSH has the advantage of requiring only small amounts of genomic DNA. It can be used with any genome, even one that has not yet been characterized. It is especially useful for identifying the large genomic differences typically found between bacterial genomes. For example, SSH identified the genetic material that causes Kaposi's sarcoma, a skin lesion associated with HIV and AIDS. At Livermore, SSH has been useful for finding differences among anthrax strains and other potential agents of bioterrorism. Comparison of *Y. pestis* and *Y. pseudotuberculosis* revealed seven DNA regions in *Y. pestis* that do not occur in *Y. pseudotuberculosis*. Four of them occur very closely to one another on the *Y. pestis* genome. "It is fair to assume that pestis acquired this region during its evolution from *Y. pseudotuberculosis*," says Radnedge.

To learn more about the function of genes in these areas, Garcia and others are beginning "*knock-out*" studies. They will inactivate, or knock out, one gene at a time and test the resulting bacterium on an animal to see how the host and its genes respond. This is slow, laborious work, but it will help to determine what the function of each *Y. pestis* gene is, if any, and what gene or genes in the host are expressed as a response. This detailed examination of pathogen–host interaction for plague will be the first of its kind.

Being Prepared

Research to date on plague lays the groundwork for additional work planned at Livermore in the areas of microbiology, proteomics (the global study of proteins), bioinformatics (the integration and analysis of biological data), and biological modeling for the NNSA's Chemical and Biological National Security Program. Some of the research will elaborate on plague, some will examine a broader spectrum of human pathogens, and some will further the development and use of biodetectors, mass spectrometry, and other technologies. In the U.S. today, *plague pops* out of the rodent population and into the human populace occasionally in the desert Southwest. It is a larger problem in a few other countries. But the real fear is that plague could be used as an agent of mass destruction. At least in industrialized countries, it is unlikely that plague would cause the huge number of deaths that occurred during earlier epidemics. Better sanitation, a more educated populace, and a far superior medical system would likely prevent that. But the world needs to be prepared.

10

ANTHRAX

A letter containing anthrax had been delivered to the Washington office of Senate Majority Leader Tom Daschle. An undetermined number of staffers in Senator Daschle's Hart Senate Office Building suite had apparently been exposed to the potentially lethal biological agent when the letter was opened. President George W Bush just minutes before had told the nation about the letter during informal remarks following an event at the White House. It shocked the public who had spent the morning hosting a roundtable on bioterrorism at the Tennessee Emergency Management Agency headquarters in Nashville. People listened carefully as about three dozen doctors, nurses, hospital administrators, firefighters, police, and other law enforcement, public health, and emergency personnel from all across the state talked about how unprepared they were for the growing threat of bioterrorism.

Doctors and media did little at the time of this initial press question into seven days that would severely test much of what we thought we knew about bioterrorism. Seven days that would challenge our fundamental clinical understanding of anthrax—how to diagnose it, how to treat it, how to protect those who may have been exposed—and how to communicate with the press and the public about a public health emergency with shifting facts and a fluid, rapidly evolving scientific knowledge base.

The coming week also would bring sharply into focus the vital role our public health system plays in responding to such an attack. After all, this was part of the first biological attack on U.S. soil involving anthrax and was only the second known biological attack in the United States in at least a century.

MONDAY

As we prepared to host the bioterrorism roundtable in Nashville, a Florida man already had died of inhalational anthrax, and an assistant to NBC News anchor Tom Brokaw had been diagnosed with skin, or cutaneous, anthrax. An additional case of each form of the infectious disease would be confirmed later in the day.

So everyone in that room knew that the threat we once considered remote was growing. Frontline responders, those who will answer that panicky phone call or first see a person with symptoms, lacked appropriate training and protective equipment. State and county public health laboratories lacked adequately trained epidemiologists and state-of-the-art facilities and equipment. Community hospitals had no system to share information with each other or with local public health facilities in a timely way.

Vanderbilt Hospital—one of Tennessee's premier teaching and referral hospitals, where we had worked for nine years as a transplant surgeon before coming to the U.S. Senate—acknowledged that it did not have a bioterrorism preparedness plan in place and had not done training exercises to deal with this emerging threat. Nor had any of the community hospitals. From rural doctors to city police and firefighters to state public health officials, all the way up to the governor's office, the story was the same: They just weren't prepared to respond to what might happen.

Following the conference, we walked into an adjacent room for a press briefing, and the question was no longer about what might happen. It was about what had already happened. When asked about the Daschle letter, we simply replied that we were unaware of the report but that it would have to be confirmed. To be honest, we minimized it a bit in my own mind because we recalled receiving an anthrax hoax letter three years before. If the report was correct, this would be the first witnessed exposure of multiple people to the release of airborne anthrax.

Our next scheduled stop was a speech about bioterrorism to the Nashville Rotary Club at a downtown restaurant. Ironically, we had picked the topic three weeks earlier when the meeting was booked, before anthrax was on anyone's radar screen. By the time we arrived at the Rotary meeting, we had been notified that my own Public Health Subcommittee staff was on the same floor in the Hart building just down the hallway from where the letter had been opened. We were told that there was apparently a large amount of a powdery substance in the letter and that nasal swab testing and distribution of a short course of antibiotics had begun for those in the Daschle suite. We sensed that the issue was escalating.

We were able to get periodic progress reports by telephone from our subcommittee staff. At this point, the Hart building remained open and people were still at work at their desks. In fact, the building would stay open for another day and a half. The risk of exposure was unknown at the time. We returned to Washington that afternoon and was asked by Senate Republican leader Trent Lott to be the liaison for the Republican senators to the fledgling medical and law enforcement investigation into the anthrax exposure at the Hart building.

Already, those within Senator Daschle's suite had been given nasal swab tests along with a three-day course of antibiotics. And a public health command room had been set up by Majority Leader Daschle in the secretary of the Senate's office on the third floor of the Capitol. It was in this room that data were reported and shared and strategies to deal with the evolving public health and environmental issues were discussed and developed.

Officials representing many agencies—the Centers for Disease Control and Prevention (CDC), the Capitol physician's office, the Defense Advanced Research Projects Agency (DARPA), the Senate sergeant at arms, the Capitol police, Senate leadership, the deputy surgeon general representing the Department of Health and Human Services, and the director of the District of Columbia health department—were almost always in that room for the next several days, coordinating among themselves and then reporting back to their home agencies. To provide information to the Senate staff and the public at large, we held two press conferences on the first full day and then scheduled daily press conferences for the next week. But many staffers were still confused and anxious about their potential health risk.

Just a few doors from Senator Daschle's suite, my own health subcommittee staff members were experiencing firsthand the difficulty in obtaining helpful information during a public health emergency. Although they knew about the letter, primarily from press reports, almost nothing else was reaching them. Like the rest of the nation outside the walls of the building, their principal source of information was TV.

On that first day, staffers from Senator Daschle's offices turned to our health subcommittee staff with questions about anthrax and their own health risk, in part because they knew that my staff had been working on issues related to bioterrorism for a long time and in part because, as we are the Senate's only doctors, they knew that our staff and we are involved in most health-related issues, Is anthrax contagious? If we have been exposed, is our family at risk? However, no official word came from those conducting the investigation until later that afternoon, when a police officer visited my staff in the Hart building to confirm that a letter possibly laced with anthrax had been delivered to Senator Daschle's office.

In response, they were closing off a section of the Hart building. The officer would remain on duty outside my staff's door, but the primary reason was to ensure that the press did not bother them as they continued to work. Staffers moved in and out of that wing of the Hart building, totally unaware of any health concerns.

The ventilation system had been shut down within an hour of the incident to avoid potential spread of anthrax, and the staff was told to expect the offices to be warmer than usual. At that time, there was no discussion about their own health risk, and previous experience would suggest that they were not in harm's way.

Nasal swabs to determine how widespread exposure to the anthrax spores had been were eventually obtained from everyone in the Hart building. On Monday, only those present in the vicinity of the Daschle suite when the letter was opened were given preventive antibiotics pending results of the nasal swab tests.

When nasal swab test results confirmed the direct exposure to anthrax of twenty-eight people inside or immediately adjacent to the Daschle suite, anxiety across Capitol Hill soared. The innocent opening of a letter, a routine task that is done millions of times every day in offices across the country, suddenly escalated into a public health crisis that truly frightened many people who work on the Hill. Congressional mail was quarantined, and a month later on November 16, 2001, a second anthrax-laced letter—this one addressed to Sen. Patrick Leahy—was discovered by government investigators.

Few of the office workers on that first day really understood what a positive nasal swab meant. What about those whose tests came back negative? Did they have anything to worry about? Why did some of them still have to take antibiotics? It seemed so confusing to everyone involved, including members of the press who were trying to interpret it all.

Anxiety was high. People were not getting the answers they needed. And it was only just beginning.

TUESDAY

Staff members who had continued working in their fifth- and sixth-floor Hart offices through Monday were surprised the next morning to find a police barricade barring the entrance to the southeast corridor. They were told that they could not return to their offices and were directed to obtain a nasal swab test and more information from the medical crew.

Hundreds and hundreds of concerned individuals lined up for the nasal swab tests in the Hart building. It was difficult for the necessary medical supplies to be maintained. Several times, the line stalled owing to lack of the antibiotic ciprofioxacin (Cipro) or the necessary sterile cotton swabs. Anyone who wanted testing simply had to show up; no one was turned away. Hundreds of people from all over Capitol Hill came to be tested, even if they were not in the Hart building. Many people showed up to be tested out of confusion, others out of fear.

Everyone receiving a nasal swab that day was given a three-day supply of Cipro and told to return for the test results; at that time they would be told whether they would have to take the antibiotics for sixty days. Because of all the confusion, misinformation floating around, and the fact that very little information was then being made available, our staff and we immediately went to work to make the official Senator Bill Frist website a central place where anyone could go to find accurate, up-to-date, pertinent information both on anthrax generally and on the rapidly evolving situation in the Senate office buildings.

It was not the first time my office had to deal with bioterrorism. Three years earlier, after we received the hoax letter labeled "*anthrax*," we addressed with our staff mail-handling protocols for any letter that appeared suspicious. We were able to go back and find these protocols and post them immediately on the website so everyone had access to the information. We were among the first sites—even before the postal service—to have practical information on what to do if you received mail suspicious for anthrax.

Working with the Capitol physician and others involved in the situation, our staff gathered information from the command-room briefings and continually updated the postings. It was obvious that people were growing anxious for information about anthrax and developments on the Hill.

They tried to get on the CDC website, but it had crashed and no information was available on it. Where could they go to get information about their own personal health risk? Were the twenty-eight people who tested positive now infected with the most deadly form of anthrax disease, inhalational anthrax? (They weren't. The test merely confirmed that they had been exposed to the spores.) Why did some people around them have positive test results while their results were negative? How long would they be out of their offices? Should they go to work the next day? People were desperate for information. And desperation can lead to frustration. And frustration can lead to anger. And they didn't know where to turn.

Wednesday

For hours after the anthrax-laced letter was opened, information was still scarce. Staff continued to show up for work, and the testing site was moved to the Russell Senate Office Building because the Hart building was closed. As thousands of Senate and House staff and those who had been visiting Senate buildings on Monday lined up for their nasal swabs and three-day Cipro supply, our staff busily printed out information on anthrax exposure that we gathered from my Senate website and others and handed it out to those who waited in line. They were understandably eager for information.

The Senate and House leadership met early Wednesday morning and received reports that environmental cultures from several locations had tested positive for anthrax, suggesting an even higher potential for contamination than previously thought. Decisions were made independently by the House and Senate leadership to close their respective office buildings to allow more comprehensive testing.

When the Senate leadership presented the decision to the full membership, though, they learned that senators felt strongly that their session on Thursday should go on as scheduled, During the meeting, we briefly addressed our colleagues, offering our views from a medical standpoint on the anthrax exposure so far. Our discussion seemed to reassure many of them.

After hearing from officials involved in the investigation, several senators made it clear that they believed Senate business should not be stopped, because it would be a sign of giving in to the terrorists. At the time, most felt that the events occurring on the Hill were likely related to the September 11 attacks. This was, many believed, round two of the assault on the nation's capital. A plane had struck the Pentagon and killed 189. Another that crashed in Pennsylvania, killing 44, almost certainly was headed for Washington. And now many thought, came deadly anthrax, aimed at the highest-ranking member of the U.S. Senate. In the end, we decided to close the Senate office buildings later that day but to stay in session in the Capitol building itself on Thursday. House members, in a separate meeting, decided to close their office buildings and to adjourn.

The press made a big deal about the House leaving and the Senate staying. It was an uncomfortable situation for both houses, with accusations flying back and forth, mostly fueled by the press. To us, this just reflected an initial miscommunication between the House and the Senate at the leadership level, in large part arising from the lack of certainty surrounding the interpretation of environmental culture results that were slowly coining back. Apparently, House and Senate leaders had left the earlier morning meeting with different understandings as to what they would do regarding closing the buildings. Policy decisions that could affect the safety of personnel had to be made on the basis of incomplete information and inadequate science. Such is the nature of bioterrorism.

At the request of the Senate leadership, we briefed the chiefs of staff of all the senators in the basement of the Capitol at 1 P.M. It was a tense meeting. The same officials who had briefed senators in the morning gave a quick update, but the high-level staffers in the room were angry, frustrated, and resentful. And appropriately (though unavoidably) so. They were responsible for their offices hey said, yet they were not getting any information to tell their staff members. Were they safe? Why were senators getting briefed early and the staffers so much later? Had they been exposed to something that could harm them? Were the clothes they were wearing safe, or could spores on the clothing be taken home and infect their children? When could they get back in their offices?

How would they get their information if they were at home with out access to their computers? Just listening and doing my best to respond made me realize how critical communication is, especially times of terror. With offices closing, the normal flow of information through e-mail was instantly cut off because people could no longer get to their office computers, and they couldn't access internal Senate information from their home computers, At the staff meeting, we announced that we would make my official personal Senate website available for the foreseeable future, posting all the information to which we had access. We gave them the website address and instructed our staff to stay on it full-time to keep it current throughout the developing situation.

The website contained basic information about anthrax as well as particular ways to deal with this recent attack, including the protocol for opening mail, easy-to-access references, frequently asked questions, and updated information about the current test results, how to obtain your test results, and where additional individuals could be tested. It became a central repository for information for staffers and senators alike.

Everyone wanted to know more about skin and inhalational anthrax, and the initial TV news reports by political figures and health officials were confusing. What does the rash look like? Even as a physician familiar with infectious diseases because of our transplant experience, we'd never seen it. So we immediately called our colleagues around the country, found pictures of the rashes, and with in a few hours posted them on the website for the world to see. As questions came throughout tile next several days, we posted them with the best answers available. What about pregnant women who were exposed? Go to the website. What are the side effects of Cipro? Go the website. You could not get that information from the family doctor because the doctor had never been taught about the symptoms and signs of anthrax disease. The answers generally are not in medical textbooks because anthrax is too rare. But we were able to get it all up quickly on a centralized website. Another lesson learned.

Because of our long-standing interest in bioterrorism and the work we had done putting together the Public Health Threats and Emergencies Act of 2000 a year previously, we already had a section on bioterrorism on our site. We were able to expand it quickly over a few hours and updated it twice a day as reports came out of the public health command room at the Senate. Our staff made sure this site was linked to other reputable sites with useful information on bioterrorism, biological agents, and public health safety, including the CDC (the overall best site) and other government and university sites. Thousands of people in Washington began to visit our site. Indeed, because it provided current and accurate information, people from around the country began to use it as a primary site.

Thankfully, other than the initial twenty-eight staffers, no one else on the Hill tested positive for exposure to anthrax. So only individuals on the fifth and sixth floors of the southeast corridor of the Hart building were initially given the sixty-day course of antibiotics (All the nasal swab tests reported for my staff were negative, although one individual's test result was lost.)

More than six thousand nasal swabs were tested through the Capitol attending physician's office alone. In addition, thousands of environmental cultures were taken over the next several days. The laboratories were stretched to their limits, and their capacity to handle a sudden rush of new test requests—their "surge capacity"—was surpassed. People were working around the clock. When test results confirmed that anthrax contamination had been detected in the Dirksen Senate Office Building mailroom and elsewhere in the Hart building, as well as in one of the House buildings, the decision was made to close all of the congressional buildings Thursday until further environmental tests could be completed.

This only added to the anxiety of several thousand staffers who were left to wonder whether they had been exposed and were at risk for anthrax infection. Paralysis began to set in.

Thursday

We held two press conferences on Thursday, still operating with the same team of public health officials out of the command center on the third floor. The CDC continued to send new people up from Atlanta, but there was crucial consistency in leadership in the command center throughout this period.

Having two press conferences separated by only six hours allowed us to answer the onslaught of questions from the media and provide updates as we continued to glean new facts. But equally important, these press conferences gave all of us the opportunity to establish who the public health command team was and to explain to the public what anthrax is, what the relative risk of exposure is, what Cipro is and what its side effects are, how anthrax is treated, and what the difference is between nasal swabs that measure exposure and other tests that measure disease.

On Thursday, the testing site was moved out of the Senate office buildings off Capitol Hill to a day care center about three blocks away. Anxious to see if this new site, identified only late the night before, could be set up that quickly to handle the hundreds, and maybe thousands, of people requesting testing and antibiotics and information, we stopped by the off-site test facility early that morning. We were impressed. We especially respected the uniformed public health officials who took the swabs, counseled anxious people, and handed out antibiotics. We had the director of our health subcommittee staff go by the site every two hours to make sure that things ran smoothly over the course of the day. The line initially wrapped around two sides of a city block. But all were calm, although most had to wait an hour in line, and things ran efficiently. The only hitch came when parents complained that we should not have allowed the name of the day care center to appear in the paper.

Friday

By Friday, it seemed as if the anthrax outbreak had been contained. Everyone who had been exposed had been identified, and all were being treated. Based on everything we knew from both a scientific and a public policy standpoint, no one would get sick. All was under control.

Communication, though imperfect, had improved. The almost hourly thirst for new culture results and test results seemed to be subsiding since most tests had been completed. People had grown accustomed to the now familiar voices of those on the Hill, including me, who stepped up to the microphones to explain what was happening.

Yet, we were still operating in a new atmosphere characterized by much uncertainty. There was so much we didn't know. How will people respond to long-term use of antibiotics? Could we have missed someone? Was it possible someone had just been visiting the Hart building on Monday and their traveled back across the now unaware that he should be on antibiotics? Could the ventilation system have spread the spores farther than we thought? And what about the Dirksen mailroom? How could it have been contaminated? Should we have paid more attention to that? The answer we soon learned, was yes. We should have.

Saturday

The calm was short-lived. On Saturday morning, we went to the capitol to attend what we thought would be a short briefing among of us on the response team and to plan additional steps for the upcoming week. At the beginning of the meeting, someone at the table mentioned that the CDC team had been contacted by a local community physician Friday evening. A patient a few miles away from Capitol Hill had presented with signs and symptoms that might possibly be consistent with the inhalation form of anthrax: shortness of breath, an abnormal chest X ray with enlarged lymph nodes but clear lung fields. But the symptoms were also consistent with the flu. No blood cultures or other tests had confirmed the presence of anthrax, but all the pertinent tests were pending. The results would not be back for another twenty-four hours.

When we heard the details, we were startled—truly alarmed. We realized then that what we'd thought would continue to be a relatively controlled situation on Capitol Hill, with all potential victims already identified and appropriately treated, might abruptly become a national emergency.

According to everything we knew to be true about anthrax at the time, it would have been impossible for someone miles away from where the anthrax-laced envelope was opened to have been infected with the disease, We had relied on our current textbook understanding of the disease: Inhalational anthrax disease does not occur unless there is direct inhalation of more than ten thousand spores.

Given that numerous witnesses saw the letter opened in the suite of offices in the Hart building, it did not seem possible that enough spores could have escaped to infect someone who was not fairly close to where the letter was opened.

What we were discovering was that even the information from the best medical scientists and public health specialists in the United States was wrong. Dead wrong. We did not know enough. No one did. We as a nation were underprepared scientifically for bioterrorism.

Although the test results would not be back for another day and the clinical data were really not very clear, we believed the information could portend a national catastrophe. Was this part of a much bigger conspiracy or terrorist activity? Would we start seeing cases around the country now? Were postal workers safe? Would our mail system be shut down locally—or possibly nationally—just as the air transportation system had been paralyzed a month previously by the terrorist attacks on Washington and New York?

What was next?

With bioterrorism, the perpetrator does not have to be present, the weapon is invisible, and the victims may not become apparent until days after exposure. It is much different from any other type of crime.

The one thing we knew was that all certainty had disappeared. We were seeing things happen that had never even been envisioned by scientific or health officials. Just as five weeks before, the September 11 terrorist attacks at the Pentagon and World Trade Centers were beyond what anyone had ever envisioned could happen.

We immediately got up, went across the room to the telephone, and called the White House. We asked to speak to Tom Ridge. Just a few weeks earlier, Ridge, the former governor of Pennsylvania, had been appointed homeland security director by President Bush governor Ridge called back immediately and we told them that we no longer had a local public health situation but a potential national emergency with disease appearing and behaving like we had seen before. We said that this could explode as a national security issue and a national public health emergency.

If people could get infected by simple exposure to mail (not the opening of mail), would we see others presenting with symptoms round the country over the next several days? Would we be prepared if they did? We asked Governor Ridge to give us a bit more time to get more information, and we agreed to have a conference call one hour later with the nation's highest officials in charge of emergency preparedness and response. Within an hour, the team at our public health command center Capitol had a conference call with key figures from around the century: Tommy Thompson, secretary of health and human services; Bill Knous, director of the Health and Human Services Office of Emergency Preparedness; Dr. David Fleming, from the CDC in Atlanta; Tom Ridge; and Joe Allbaugh from the Federal Emergency Management Administration.

We opened the conference call and asked each of the members of the command team in the Capitol meeting room to speak for approximately five minutes. The goal was to share all that we had learned over the previous week on Capitol Hill with the federal officials on the conference call, who clearly would soon need to become more involved in the surveillance and response. It was clearly stated that significantly more federal resources would be necessary to handle the increasing burdens placed by more extensive environmental testing and decontamination. The officials heard how the thousands of swab tests performed to date had stretched beyond the surge capacity of our public health system.

At the end of the conference call, we were certain that everyone on the phone understood the potential importance of the critical new suggestion that someone outside the immediate system had been infected. And they had a clear picture that our local resources had to be reinforced immediately with federal resources. In responding to bioterrorism, this vertical communication from local authorities directly to federal officials is vital. Secretary Thompson closed the call, telling us to make a list of everything we would need. He assured us that it would be provided.

SUNDAY

Sunday morning at about ten o'clock, just before we were to go on one of the Sunday talk shows to discuss the developments on the

Hill, we learned that the patient's test results had come back. The diagnosis was inhalational anthrax. The patient, we learned, worked at the Brentwood post office in Washington D.C. Immediately, the local public health office, which had fortunately been a part of the public health command team from day one, fully mobilized and within hours began testing employees from the postal facility at a site made available by Washington's mayor, Anthony Williams.

That afternoon, we dropped by the District of Columbia office building where the testing was conducted. Again, it was impressive seeing groups of about twenty people at a time being moved from station to station, receiving information, counseling, testing, and antibiotics. The mood overall was pretty good. There was order. No panic. No anger.

Then again, no one had died—yet. But even as the testing was going on, another postal worker had entered the hospital, apparently with anthrax, Within two days, postal worker Thomas Morris Jr., fifty-five, had died. Later, a second postal worker, Joseph Curseen Jr., forty-seven, passed away.

These were tragic deaths that no one could have foreseen, knowing what we did at the time. We were basing all of our actions on specific clinical assumptions, the best information that could be found in the most current textbooks and medical literature, about a bacterium that we had been studying for centuries. We knew all identifying clinical features of anthrax—swelling of the lymph nodes of the middle of the chest and flu-like symptoms—but our epidemiological understanding—our understanding of how the disease spreads—was inadequate.

And no one had realized it.

Our assumptions had been based principally on how the anthrax bacterium acted in settings that were almost pre-industrial—before buildings were air conditioned, before technology allowed us to sort mail with the force of air, and before we had advanced medical technologies to help us stabilize patients and make more definitive diagnoses. For the most part, what we knew about anthrax based on how it occurred naturally—not on how it could be terrorists with the intent to kill and to terrorize. These are new times.

We had to reexamine our assumptions about anthrax—about the likelihood of contracting inhalational anthrax compared to cutaneous anthrax, about the relative risk to those who had little or no contact with the actual anthrax-laced letter, and about the best ways to diagnose and treat those with the disease.

As with all areas of medicine, the body of knowledge is an ever-changing enterprise, updated daily with more information from various studies. However, we had not updated our understanding of anthrax or other potential biological agents in years, primarily because of a lack of data. In short, we weren't ready for what occurred.

Once the impact of anthrax spread beyond Capitol Hill and into the U.S. postal system, spokespersons for this national issue were the federal officials charged with responding to this threat. We continued answering questions on television shows and the news, providing information about our response to this particular biological attack as well as to any future attacks.

Given that my website received as many as forty thousand hits a day, it was clear that people were hungry for knowledge. Generally, as I have discussed bioterrorism in recent months, people ask many of the same questions. We have noticed that when people get timely, straightforward answers to their questions, they become more relaxed and reassured. And that's why the present title was written—to answer the questions most frequently asked, to discuss some of the questions you may have thought about but never asked, to help reassure and inform American families about bioterrorism.

The purpose is to provide information that's as reliable and accurate as possible, to as many people as possible, in these times of greater-than-normal anxiety. Our nation is on high alert. We are being faced with new challenges, in our places of work and in our homes. But our nation, our officials at all levels of government, and the American people are taking the steps necessary now to be prepared to meet whatever challenge comes our way.

Toward this end there are things that each and every one of us can do. One of the most powerful is to become knowledgeable and informed. An understanding of some of the basics on how to prepare for, and respond to, the use of microbes as weapons goes a long way to reduce anxiety and minimize any chance of paralysis in our lives.

In the war against bioterrorism, information is power.

Assessing Risk

The anthrax bacterium grows naturally throughout the world. There are about two thousand labs in the United States alone that hove anthrax samples, not to mention hostile foreign nations such as Iraq that have large supplies. However obtaining a strain that would do the greatest harm would not be simple. *Anthrax* spares are highly resistant to drying and to extremes of heat and cold. The spores also are resistant to some disinfectants. Spores can survive in the environment for decades. It takes technical skill to refine anthrax to the extremely finy size required to get into the lungs, the staging ground from which it launches its often deadly attack on the body. But as we have learned, someone with that skill can send anthrax though the mail with deadly consequences. Considerably more sophistication would be required to manufacture anthrax for an aerosol release that could inflict moss casualties. It was thought before the postal attacks that inhalational anthrax would be fatal in 80 to 95 percent of cases. However, six of the eleven people who contracted the disease last fall survived thanks to fast treatment with antibiotics. Although most Americans knew little about anthrax until recently, the disease has been around at least since biblical times.

Many believe that anthrax fits the description of the fifth *"grievous" plague* that killed the Egyptians' livestock after Pharaoh refused to free the Israelites, as recounted in the book of Exodus. Descriptions of similar plagues can be found in the writings of Homer, Virgil, and the ancient Hindus.

But it was the use of the mail to send lethal, *anthrax-laced letters* to media outlets and government offices last fall that suddenly made this ancient disease a household word. Even now, months after the postal attacks that infected eighteen and killed five, there are a lot of misconceptions about anthrax. For instance, a poll conducted more than a month after the attacks found that a quarter of Americans believed the disease is *contagious*. It's not. But it is no wonder people are confused. Much of what we know about anthrax as a *biological weapon* was learned during the attacks last fall.

What is Anthrax?

Anthrax is an infectious disease that can affect both animals and humans. The illness is caused by a spore-forming bacterium called *Bacillus anthracis*. Three forms of anthrax can occur: *skin* (or in medical terms, cutaneous), *inhalational*, and *gastrointestinal*. The form varies based on how the anthrax bacteria enter the body.

The bacterium gets its name from the Greek word for coal, *anthrakis*, because the skin form of the disease—which is by far the most common in humans—is characterized by skin sores that turn coal-black.

Bacillus anthracis can switch back and forth between two states: the active "*vegetative*" form, and the dormant "*spore*" form that you have heard so much about. The bacterium is only technically alive when it is in the vegetative state. That's when it can reproduce, take in nutrients, and get rid of wastes. When the bacterium senses a lack of nutrients or water, it dries out and encases itself in a thick, hard shell called a spore that protects it against extreme heat, cold, and even some types of radiation (but not the gamma irradiation currently being used to treat the mail).

In this state of suspended animation, the spores can survive in the soil for decades. That's why anthrax primarily strikes grass-eating animals, such as cattle, goats, sheep, and horses. In fact, viable anthrax spores can still be found along the cattle trails of the Old West.

Where is Anthrax Found

Anthrax is not uncommon in agricultural areas of the world, such as Africa and the Middle East. An estimated 20,000 to 100,000 human cases occur globally each year, most of which are skin infections. However, anthrax is rare in industrialized nations such as the United States.

Between 1944 and 1994, just 224 cases of the skin form of the disease were reported in the United States. In the last century, only 18 cases of inhalational anthrax—the most deadly form of the disease—were reported in this country. Two of those cases were related to laboratory work, and not a single inhalational case was reported in the twenty-five years leading up to September 11, 2001.

For thousands of years, the threat of anthrax to humans was primarily limited to farm-workers, woolsorters, and, in rare cases, those who ate tainted meat. That all changed with the postal attacks that followed in the wake of the September 11 terrorist attacks. As of January 1, 2002, 18 newly confirmed cases of anthrax had been identified in the eastern United States and 5 people had died of the disease. This recent outbreak makes it abundantly clear that anthrax can be used as a bioterrorist weapon against civilian populations.

How is Anthrax Spread?

It is important to note that anthrax is not a contagious disease. That means it can't be passed from one person to another. The only way to be infected is to come into direct contact with anthrax spores through one of the ways outlined below.

Humans can contract anthrax several different ways: by directly handling infected animals or contaminated animal products (such as wool or hides); by inhaling anthrax spores into the lungs; and by eating undercooked meat from infected animals. The recent cases from tainted mail demonstrate that infection can also be acquired by handling artificially contaminated items, such as letters or packages.

Handling contaminated animals, animal products, or artificially contaminated items can lead to skin infections. Anthrax spores enter the skin through minor cuts or scrapes; the most frequently infected areas of the body are the arms, hands, face, and neck—areas most often left exposed. The average time from exposure to onset of illness (the *incubation period*) for skin infection is one to seven days. *Cutaneous anthrax* accounts for an overwhelming majority (an estimated 95 percent) of all anthrax cases.

Inhaling the spores can lead to *inhalational anthrax*. This is the form that concerns us most, because it is the form most likely to be used as a biological weapon and it is the most deadly. Experts believe that the average lethal dose for inhalational anthrax is ten thousand spores, although in view of the recent postal attacks, we now believe that a smaller number can be fatal, especially for the elderly and those with a weakened immune system.

Once inhaled, the microscopically tiny spores each one less than more one-twentieth the diameter of a human hair—lodge in the lungs and then spread to lymph nodes in the middle of the chest between the lungs, in what is known as the mediastinum ,The spores convert back to their active vegetative state and begin multiplying with a vengeance.

In hours, or perhaps a few days, a *deadly toxin* is released that spills over into the bloodstream. Severe shock and death frequently follow. Antibiotics are most successful if begun before the toxins are released.

In the inhalational anthrax cases following September 11, the average time from exposure to the bacteria to the onset of symptoms was four days. But experts believe that illness may occur as long as sixty days after exposure to anthrax spores, because observations have shown that the spores can take that long to change to active bacteria. That explains why preventive (*prophylactic*) antibiotics are typically given for a sixty-day period. Nonhuman primate studies have suggested that illness may occur up to a hundred days after exposure. This is why after very high exposure to spores, an additional for days of preventive antibiotics may be recommended.

The third, and least common, way to get anthrax disease is by eating contaminated meat from infected animals. This type of exposure leads to *gastrointestinal anthrax*, which usually has an incubation period of two to five days.

Symptoms of Anthrax

Symptoms of anthrax depend on how the disease is contracted. Here what to look for in each case.

Skin (cutaneous) anthrax

Cutaneous anthrax begins as an itchy bump that resembles an insect bite. Within a day or two, it grows into a round, fluid-filled ulcer about a half-inch to a little more than an inch in diameter. A depressed, painless black scab—the characteristic feature of the disease—then forms in the middle of the ulcer. The scab loosens and falls off within one to two weeks, leaving only minimal scarring, if any. Severe swelling may occur around the ulcer and painful swollen lymph nodes may develop in the area. If left untreated, an estimated 20 percent of people will die from cutaneous anthrax. However; with proper antibiotic treatment, the odds of survival are extremely high.

Inhalational anthrax

The most deadly form of the disease progresses in two distinct stages. The first phase resembles a flu-like illness and is marked by fever, chills, headache, nausea, vomiting, muscle aches, and fatigue. Within hours to a few days, during which some patients experience what appears to be a brief period of recover the disease moves into its more deadly second stage. This is characterized by high fever; fluid in the lungs, severe breathing problems, and low blood pressure. Massive swelling of lymph nodes in the chest may occur. About half of those with the inhalational form of the disease develop anthrax meningitis if untreated. Left untreated, up to 90 percent of patients may die. However, with early treatment using appropriate antibiotics, the risk of death may be much lower (30 percent). Exciting research on new antitoxin therapy will likely reduce this risk much further in the future.

Gastrointestinal anthrax

Gastrointestinal anthrax can cause two distinct types of illness. The more common form affects the small or large intestine and resembles a severe case of food poisoning. Initial symptoms include nausea, loss of appetite, vomiting, and fever. These symptoms are quickly followed by abdominal pain, vomiting of blood, and severe diarrhea. A less common form of gastrointestinal anthrax causes a severe sore throat, fever; trouble swallowing, and, sometimes, ulcers in the mouth or back of the throat. Very tender swollen lymph nodes may occur in the neck. If gastrointestinal anthrax is left untreated, about 50 percent of patients will die from the infection.

Anthrax is Aerosolized

That simply means it has been made airborne. Anthrax spores naturally tend to clump together in chunks so big that your body's natural mechanical defense system would intercept them before they got deep into the lungs where they cause the biggest problems. So someone would not only have to be able to refine the particles to microscopically small sizes—an estimated one to five microns in width—but also have enough knowledge to use chemicals to break up the clumps. During this very difficult process, the bacteria can often be rendered inactive.

Weaponized Anthrax

Aerosolizing anthrax is one type of "*weaponization*," a word that refers to engineering the anthrax spores, either physically or genetically, so they are delivered more efficiently and are more deadly. In the cases following September 11, there was evidence that the anthrax spores had been specially treated so they would remain suspended in the air for prolonged periods, making them more likely to be inhaled because they could literally float out of an envelope.

Another type of weaponization of anthrax, perfected by the Russians in the 1970s, is to engineer it to make it resistant to antibiotics. This fortunately was not the case with the anthrax sent last fall. Given antibiotics' effectiveness against anthrax when used early enough, this is the largest bioterror concern. Other types of weaponization include special milling to refine particle size, adding agents such as stabilizers or skin irritants, and genetic modification to alter the incubation period.

Most experts believe that only a person, a group, or a country with access to advanced biotechnology would be capable of manufacturing and delivering a lethal anthrax aerosol. For example, the Japanese terrorist group Aum Shinrikyo, which was responsible for the release of deadly Sarin gas in the Tokyo subway system in 1995, had previously tried to disperse aerosolized anthrax and botulism throughout Tokyo on several occasions. For reasons that remain unclear, the attacks caused no illnesses.

Anthrax as a Weapon of Mass Destruction

The answer is yes, though to date it has not been. Fortunately, it's far more difficult to convert anthrax into a weapon of mass destruction than you may have been led to believe. First, only certain strains of anthrax bacteria are exceptionally deadly. A bioterrorist would have to have access to a particularly virulent strain and then brew a large batch of the microbes. The bacteria would have to be dried and converted to spores, then refined into very, very small particles.

The recent distribution of anthrax through the mail system infected at least eighteen people and killed five. The mail system was paralyzed regionally, Congress was essentially shut down for four days, buildings were closed for months. The country was terrorized. But anthrax was not used, as it might have been, as a true weapon of mass destruction. The same amount of anthrax placed in the ventilation system of a building could have exposed thousands to a lethal dose.

For more than three decades, scientific, military, and health experts have tried to analyze the consequences of a large-scale anthrax attack. The worst-case scenario would be that some nation or group was able to spread anthrax from an airplane over a major metropolitan area. In an analysis that is over thirty years old and conducted long before we developed the National Pharmaceutical Stockpile and early-mobilization program, the World Health Organization estimated in 1970 that the release of aerosolized anthrax over a densely populated area with 5 million people could result in 250,000 casualties, 100,000 of whom would die unless treated.

Anthrax as a Bioweapon

Research on anthrax as a biological weapon dates back to World War I, when the Germans tried to us it to disrupt the Allies' horse- and reindeer drawn supply lines across northern Norway. Eighty years later, scientists discovered that a lump of sugar laced with anthrax by a German spy still contained living spores.

During World War II, the United States, fearful that Japan and Germany were making *bioweapops*, first began experimenting with anthrax and other germ warfare. It was later discovered that Japanese scientists subjected Chinese prisoners of war to horrifying experiments with lethal bio-agents as anthrax, cholera, typhoid, and plague. As many as ten thousand were killed.

The British also conducted anthrax experiments during World War II, detonating explosive shells filled with anthrax spares anon island off the coast of Scotland The spores were still viable thirty-six years later, and it took an intensive, eighty year decontamination effort requiring 280 tons of formaldehyde and 2,000 tons of seawater to clean up the island. During the *Cold War*, the United States and the Soviet Union both undertook extensive bioweapons campaigns that included anthrax. But in 1969, President Richard M. Nixon ended the U.S. offensive biological warfare program and ordered all stockpiled weapons destroyed. In 1972, the United States and more than a hundred nations including the Soviet Union, Iraq, and Iran signed the Biological and Toxin Weapons Convention, banning bioweapons.

The Soviets, however, ignored the treaty and geared up their offensive, attack-oriented program. Over seven thousand scientists were involved in this effort. In 1979, anthrax spores were accidentally released into the atmosphere from a secret Soviet military facility in Sverdlovsk. At least seventy-seven people downwind were infected, and sixty-six died.

In the mid 1980s, Iraq launched a bioweapops program and had managed to develop weaponize anthrax by the time of the Gulf War. Iraq reportedly even tested crop-dusting equipment to spread anthrax over a wide area but could not get it to work. In another analysis, the U.S. congressional *Office of Technology Assessment* (OTA) estimated in 1993 that releasing aerosolized anthrax over Washington, D.C., could result in 130,000 to 3 million deaths—an attack as deadly as a hydrogen bomb. In addition to the horrifying human toll, the economic impact would also be devastating. The Centers for Disease Control and Prevention (CDC) has estimated it would cost the U.S. economy $26.2 billion for every 100,000 people exposed to anthrax. But, remember, most of these modeling exercises involve assumptions that might not be applicable today. For example, the OTA assessment assumed the attack would not be recognized for six days.

As you'll recall, the government grounded crop dusters for several days in the wake of the September 11 attacks after it was determined that the terrorists had asked questions of a Florida operator. We know that Iraq tested crop-dusting equipment to spray anthrax before the Persian Gulf War; but the effort was unsuccessful. Major modifications would be required to retrofit a plane's sprayer nozzles to spread anthrax in the extremely small, dry particles that would do the most harm.

Again, it would require considerable technical expertise to do this. Dosage and dispersal would be affected by atmospheric conditions, wind, terrain, sun, and other environmental factors. Nevertheless, the impact in terms of terror would be huge.

Difference between Flu and Inhalational Anthrax

The initial stage of *inhalational anthrax* does resemble the *flu* (or influenza-like illnesses), but remember, in the winter months, the flu, like the common cold, is a common illness, whereas anthrax is rare and will occur only if you are exposed to the spores. Assume you have the flu unless:

1. Anthrax exposure or cases are reported in year community.
2. You have reason to believe you have been exposed to anthrax through a suspicious letter or package.
3. People around you are suddenly and for no apparent reason coming down with what seems to be the flu or suspicious skin sores. But remember, a cluster of finlike cases is still probably the flu.

Your doctor can give you a quick test for influenza, if you are concerned, though it is not anywhere near 100 percent accurate. That said, there are some subtle differences between anthrax and the flu. Nasal and sinus congestion, sore throat, headache, and a runny nose are common with the flu and the common cold, but not with anthrax. Chest discomfort, vomiting, and shortness of breath are usually associated with anthrax, but not with the flu.

A chest X ray may also be helpful in distinguishing the two. Each of the ten initial patients confirmed to have inhalational anthrax as a result of the recent postal attacks had X-ray abnormalities such as widening of the mediastinum (lymph nodes between the lungs) and fluid around the lungs. You don't typically see these findings with the flu. It is wise to get a flu shot each year to minimize your chances of catching the flu, and thus lessening the opportunity to even have to think about whether your symptoms might be related to anthrax.

Treatment of Anthrax

Antibiotics can be quite successful in treating patients who are ill with anthrax. People who have *cutaneous anthrax* or *gastrointestinal anthrax* usually can be treated without problems. For antibiotics to be effective in treating *inhalational anthrax*, it is best if they are given early in the course of illness. Also, the CDC recommends that at least two different antibiotics be used to treat inhalational anthrax.

Antibiotics also can be used to prevent illness after inhalational exposure to anthrax spores. You have probably heard a lot about Cipro (*ciprofloxacin*) being the top antibiotic choice for preventing anthrax. But doxycycline is equally effective, if the strain of anthrax bacteria involved is not resistant to it. In fact, as the number of people initially taking ciprofloxacin during the recent postal-related outbreak neared thirty thousand, the CDC switched its preferred recommendation to doxycycline. The change was made in part because of concern that other bacteria would develop resistance to ciprofloxacin with so many people using it.

Doxycycline—which is available in generic form and is about one-tenth the cost of *Cipro*—was a good alternative under those circumstances. Although recommendations may be modified over the coming months, currently, when no information is available about whether the implicated strain of anthrax bacteria is especially susceptible to any particular antibiotic, ciprofloxacin or doxycycline is recommended for adults and children, although the course for children varies slightly. Provided the particular strain of anthrax is not resistant to it, *penicillin* is recommended by the CDC as another option for the sixty-day course of antibiotics after exposure. However, it is not considered as effective as Cipro or doxycycline for treatment of the disease. The CDC recommends that preventive treatment with antibiotics continue for sixty days, since it has been shown that the incubation period can be that long. Side effects of treatment with Cipro seen after the anthrax exposure last fall included joint aches and gastrointestinal discomfort in adults. Approximately nine thousand people in Florida, Washington, D.C., New York, and New Jersey were advised to take antibiotics for the full sixty days.

Assessment of side effects and how many people continued the entire sixty-day regimen is under way. And public health experts continue to evaluate how to deal most effectively with anthrax. It may be that, for those individuals with high exposure to airborne anthrax spores, the antibiotic regimen should be extended an additional forty days, just to be on the safe side. And for those at greatest risk of inhalational anthrax, such as those in the room when an anthrax-laden letter is opened, the option of vaccination in addition to antibiotics should be considered. In response to the events in the Hart Senate Office Building, the CDC in late December of 2001 elected to make anthrax immunization available to the seventy individuals in the immediate vicinity of where the letter was opened.

Currently, the *Food and Drug Administration* (FDA) has indicated that ciprofloxacin penicillin C procaine, and doxycycline are approved for preventive treatment following inhalational exposure to anthrax spores. In addition, *tetracycline*, *minocycline*, *oxytetra cycline*, *demeclocycline*, and *penicillin C potassium* are approved by the FDA for treatment of patients who are clinically ill with anthrax infection.

Every emergency room has plenty of antibiotics on hand for those who need them. Currently, the National Pharmaceutical Stockpile has enough antibiotics to fully treat two million people after an anthrax exposure, and recent federal funding will soon increase that number to millions more. These "*push packs*" of antibiotics can be made available anywhere in the country within twelve hours.

Treatment for Children

For children, the CDC says *Cipro* and *doxycycline* can be used for the first two to three weeks of treatment to prevent *inhalational anthrax*, and for the first one to seven days of treatment for *cutaneous anthrax*. The rest of the sixty-day course can be completed with *amoxicillin*, a form of *penicillin* that is known for its safety with children and infants, if the organism is sensitive to it. This strategy could avoid potential side effects in children.

The American Academy of Pediatrics generally recommends that *doxycycline* not be used in children under nine years old because the drug may retard skeletal growth in infants and cause discoloured teeth in infants and children. *Cipro* is not generally recommended for children under the age of sixteen because it may cause temporary joint disease in a small number of children. However; those potential side effects are greatly outweighed by the serious risk anthrax poses.

Side Effects of Cipro

Cipro is generally well tolerated. Side effects include vomiting, diarrhea, sun sensitivity, rash, headaches, and dizziness. With prolonged courses of antibiotics, joint aches have been reported. Blurred vision, hypertension, and other nervous system symptoms occur in fewer than 1 percent of patients. Caffeine and medications containing theophylline may accentuate the symptoms. Each antibiotic has different side effects. Whenever you are prescribed an antibiotic, take time to talk to your physician about the side effects of the specific medicine you are given.

Diagnosis

Simply put, there is no screening test that determines exposure to anthrax. We must distinguish between exposure and infection. Exposure is determined by environmental tests; *potential exposure* is treated with preventive antibiotics. Exposure is not infection. Infection means you have the clinical disease and you need aggressive treatment with antibiotics. More than six thousand *nasal swab* tests that were done on those who were potentially exposed to anthrax spores on Capitol Hill. These tests are best thought of as environmental tests, because all they do is tell where spores have been. They do not detect the presence or absence of disease in an individual. They are not like the more familiar throat- culture swabs taken for suspected strep throat, which are useful for making treatment decisions for an individual because they detect whether or not that person has a disease.

Nasal swabs are typically used to determine how far spores have traveled in a specific room or building where the presence of anthrax is suspected or has already been established by environmental sampling. They are useful only for public health officials to define the perimeter of potential exposure.

At the current time, recommendations for preventive antibiotics are made on the basis of probable exposure. So no one should request a nasal swab thinking it will change his or her personal care. To determine actual infection or disease (not just exposure), there is a blood culture test that's accurate and definitive. It involves placing a blood sample in a culture of nutrients and then waiting twenty-four to seventy-two hours to see if an anthrax colony grows. The lack of a fast, accurate, simple test to determine infection is one of the challenges we face. Promising new technology is on the horizon, however.

Scientists are developing a test that would use DNA technology similar to that used in criminal investigations and genetic tests to deliver accurate results within just two to three hours. And there is some indication that radio imaging used to detect infections may be able to be adapted so it picks up signs of anthrax infections before they spread to the bloodstream. Best of all, the test would take only forty-five minutes. As noted earlier early diagnosis and treatment—ideally before onset of symptoms—may play an important role in saving the lives of those who contract inhalational anthrax.

Recommended Antibiotics

To begin with, the antibiotics used to treat anthrax carry serious side effort for some people. There is no reason to subject yourself to them unless there is a real risk of disease. Even more important, if you take antibiotics when you don't actually need them, there is a chance they won't work if and when you do. Bacteria are cagey, versatile, and always changing. When exposed to antibiotics for prolonged periods, the bacteria self-select and become resistant to that antibiotic and even to related antibiotics. That creates a very real potential danger for you and for others because there may be no treatment for future disease caused by that self-selected, resistant bacteria.

Stockpile Antibiotics

It is not a good idea to stockpile antibiotics. They have a specific shelf life, and it is possible the antibiotics would expire before you needed them. Plus, if many people hoard the antibiotics, it creates the possibility that there could bc a shortage if they actually were needed somewhere. Many would be tempted to take the drugs unnecessarily, opening up the potential for side effects and the Pandora's box of potentially resistant, deadly bacteria—not only for themselves, but for society as a whole.

Vaccination for Anthrax

There is an *anthrax vaccine*, which, according to the CDC, is recommended for people aged eighteen years and older who have a high likelihood of coming into contact with anthrax spores. These include military personnel; certain laboratory workers; people whose jobs may expose them to anthrax, such as those who work with animal products from areas of the world where disease in animals is common; and veterinarians who may handle potentially infected animals (again, in areas of the world where anthrax is common). In addition, last December for the first time, the vaccine was offered to those who had high direct exposure to the anthrax in our mail system.

The vaccine is in extremely short supply, and it is not available to the general public. Currently, only one small company in Michigan is licensed to produce the vaccine. But a large-scale vaccination program really is not called for, considering the enormous cost and difficult logistics, coupled with the low probability of an attack in any given community. Remember, even with the postal attacks last fall, the odds of any one person contracting anthrax are much less than those of getting struck by lightning or attacked by a shark.

Also, the current vaccine is given in a series of six shots over eighteen months, and a yearly booster shot is also recommended. It is known to protect against skin infection and is believed to be effective against inhaled spores as well. About a third of those who get the vaccine experience tenderness and redness in the area of the shot. More severe reactions are infrequent.

Promising new vaccines are being developed, and one of our top priorities in safeguarding our nation against possible anthrax attacks is to develop, manufacture, and stockpile a new generation of vaccine. Last December, the CDC acquired 220,000 doses from the Pentagon to investigate how best to use the cur rent vaccine in the meantime. Vaccinating all the people considered potentially at risk today for occupational reasons would require at least 28,000 doses.

Opening Mail

Until as recently as October 2001, experts believed that there was little threat posed by letters coming in contact with anthrax-laced mail or going through postal machinery that handled contaminated letters. But the mysterious anthrax-related deaths of an elderly woman in Connecticut and another woman in New York in cases seemingly unconnected to the postal attacks forced them to reevaluate that stance. The probe showed that other mail containing faint traces of anthrax had been delivered to addresses near both victims, leading investigators to believe that cross-contaminated mail may have been delivered to the victims. Only trace amounts of anthrax, much less than experts believed would be needed to harm people, were found on the other letters.

Even if cross-contaminated mail was responsible for the two deaths, it is important to remember that tens of thousands of letters went through the same postal sorting machines at about the same time without causing illness. It is now believed that the elderly and people with weakened immune systems may be at a higher risk from cross-contaminated mail, But it is important to keep in mind that the risk is still very, very low. A little caution will go a long way here. There is no need to wear a mask and glove to your personal mail. If you are concerned about your personal mail, the CDC offers these simple recommendations:

1. Avoid holding letters to your nose or sniffing them before opening.
2. Don't shake or jostle the contents.
3. Wash your hands thoroughly after handling mail.

Suspicious Package

A letter or package should be considered suspicious if it is not addressed to a specific person; is marked with such restrictions as "*Personal*," "*Confidential*," or "*Do not X-ray*"; or is postmarked from a city or state that doesn't match the return address. In addition, exercise caution if the letter or package has excessive postage, misspellings of common words, a strange return address or no return address, a handwritten or poorly typed address, or incorrect titles or a title without a name.

Other suspicious signs include a powdery substance felt through or appearing on the package or envelope, oily stains, discolourations, or odor, a lopsided or uneven envelope, or excessive packaging material, such as masking tape or string. If a package or letter appears suspicious, don't open it, shake it, or carry it around to show others. Simply put it down on a stable surface. Do not sniff it, touch it, or look too closely at it. Immediately leave the area and make sure anyone else present leaves, too. Shut the door and immediately turn off the ventilation system. Ensure that all persons who touched the piece of mail wash their hands with soap and water immediately or as soon as possible, and call 911 (or local law enforcement officials). If feasible, place all items worn when you were in contact with the suspected mail piece in plastic bags and have them available for law enforcement agents. As soon as practical, shower with soap and water.

If possible, create a list of persons who were in the room or area when the suspicious letter or package was recognized and a list of persons who may also have handled the package or letter. Give the list to both the local public health authorities and law enforcement officials.

Anthrax Detection

Testing environmental samples—taken from furniture, mailboxes, and ventilation ducts, for example—is complex and painstaking. It is also very accurate and critically important to determine where and how far anthrax has spread. At the height of the postal-related outbreak last fall, the CDC looked at more than twenty-five hundred specimens, not just from New York and Washington, but from all over the country. The CDC is to be commended for rising to the challenge. It usually has about ten scientists working on anthrax-related issues, but in response to the mail attacks, it had eighty scientists working seven days a week on the case. It also provided laboratory support to field sites in Florida, New York, Washington, D.C., and New Jersey.

Typically, specimens such as envelopes suspected of anthrax contamination are taken to a biohazard lab, where they are opened under a special hood that sucks air away from the technicians and through powerful filters to keep spores from becoming airborne. The samples are then dissolved or soaked in water, and testing, which includes microscopic examination, culture, and genetic analysis, begins. For safety reasons, technicians working with the samples wear protective gloves, special disposable waterproof gowns, and N95 masks, which filter out tiny particles.

"Space Suits"

They are wearing disposable protective clothing that primarily protects the skin but also eliminates the likelihood of transferring contamination to other sites. Their respiratory devices are powered, air-purifying respirators, with full-face masks that allow them a full field of vision and are equipped with high-efficiency particulate air (HEPA) filters. Wearing a powered, air-purifying respirator with a full-face mask that has been specifically fitted to the wearer will reduce inhalation exposures by 98 percent. The disposable gloves, which are made of lightweight nitrile or vinyl, provide protection but preserve dexterity. A thin cotton glove is frequently worn under the disposable glove to protect against skin rash, which can occur when hands naturally perspire while inside gloves for prolonged periods.

Decontamination

There is no established protocol, or detailed procedure, for decontamination of buildings, and indeed the effort has proved far more complex than anyone would have anticipated. The anthrax release in the Hart Senate Office Building in Washington that I chronicled in the opening chapter provides the best example.

The building remained closed for almost three months as the best scientists and environmental experts in the world oversaw the cleanup. The entire building was closed on October 17, after twenty-eight workers tested positive for exposure. More than half of all senators had to relocate to temporary offices. On December 1, six weeks after the incident, Sen. Thomas Daschle's suite, where the anthrax-laced letter was opened, was sealed and filled with a high concentration of chlorine dioxide gas. When traces of anthrax were found in 9 of 377 environmental samples taken afterward, the Environmental Protection Agency decided to go back in and fumigate again.

This time, they used the chlorine dioxide gas in the ventilation system in those sections of the building where traces of anthrax were found and the liquid form of chlorine dioxide in the office suite itself. Afterward, a second chemical, sodium bisulfite, was used to break down the gas. One of the problems in deciding when a contaminated building is ready to reopen is that there is no agreed-upon standard of how many anthrax spores constitute a health threat. Computers, files, and books from the suite were packed up and taken to a company in Richmond, Virginia, for treatment with ethylene oxide, which is commonly used to sanitize medical instruments. In eleven other senators' offices in the building, liquid and foam forms of the chlorine dioxide and particle-filtering vacuums were used.

11

SMALLPOX

The only authorized stores of the smallpox virus are contained in two laboratories in the United States and Russia. But it's suspected that other nations, possibly including Iraq and North Korea, may have illegal supplies. Smallpox virus is stable in aerosol form, especially if weather conditions are dry and cool. It would be difficult to process smallpox virus for on aerosalized release in public, but it could be done. And if terrorists are willing to sacrifice their own lives, it is possible—although unlikely—that they could deliberately infect themselves and then spread the disease in a crowded area. Once an outbreak occurs, the disease can spread quickly throughout die world. Before the disease was eradicated last century, about 30 percent of those who caught the most severe form of smallpox, variola major, died. Getting exposed people vaccinated within four days following exposure to the virus will prevent many infections and at least decrease the likelihood of a life-threatening illness.

One of the greatest medical and public health accomplishments of the past century was the eradication of smallpox from the world. That means most Americans have never had to live with the threat of this devastating disease. But in recent years, it has reemerged in a new guise: as the scariest bioterrorism nightmare. The reintroduction of smallpox into today's world could be catastrophic. While the odds are remote that it could happen, we must be prepared. The stakes are simply too high.

WHAT IS SMALLPOX?

Smallpox is an infection caused by the *Variola* virus. The virus comes in two forms: *Variola major* and *Variola minor*. Smallpox caused by *Variola major* often is fatal; in the last century alone, smallpox claimed the lives of five hundred million people. Even as recently as the 1950s, an estimated fifty million cases of smallpox occurred in the world each year. But, in probably the most successful public health effort ever, smallpox disease has been eradicated from the face of the earth! There is no smallpox disease—anywhere.

Smallpox has been recognized as a scourge for thousands of years. Ancient writings describe a disease that resembles smallpox as early as 1122 B.C. in China, and it also appears in ancient Sanskrit texts from India. Pharaoh Ramses V apparently died of smallpox in 1157 B.C. The disease swept through Japan and Korea, reaching Europe in A.D. 710.

The explorer Hernando Cartes brought smallpox to the Americas in 1520, and 3.5 million Aztecs died in the next two years. In the eighteenth century, smallpox reached epidemic proportions in Europe, killing every seventh child born in Russia and every tenth child born in Sweden and France. Even royalty offered no protection against the disease, which claimed the lives of Queen Mary II of England, Emperor Joseph I of Austria, King Louis I of Spain, Czar Peter II of Russia, Queen Ulrika Eleonora of Sweden, and King Louis XV of France.

In 1798, however, an English country doctor named Edward Jenner proved that inoculation with a related but nonfatal virus called *cowpox* could protect against *smallpox*, offering hope for the first time that the disease could be stemmed. As vaccination became more widespread, the tide began to turn. Eventually, the cowpox vaccine was replaced by one made with the vaccinia virus, which produced fewer side effects.

In 1967, the World Health Organization (WHO) launched an ambitious program to wipe out smallpox. Mass vaccination campaigns were carried out throughout the world, and outbreaks were carefully contained. Through famine, flood, cholera, and civil war, the vaccination program pushed forward until victory was won.

The last naturally occurring case of smallpox was recorded in October 1977 in Somalia. In 1980, the WHO declared that naturally occurring smallpox had been eradicated and recommended a worldwide end to vaccinations. The last smallpox case in the United States was in 1949, and vaccinations here were halted in 1972.

How is Smallpox Spread?

Smallpox is spread primarily through respiratory droplets, or aerosols, from infected patients. A person generally becomes infectious, or capable of spreading the disease to others, once the rash starts. And smallpox patients are most infectious during the first week of illness, when lesions in the mouth ulcerate, releasing substantial amounts of the virus into the saliva.

Those with smallpox remain infectious until all of the lesions on their bodies have scabbed over and the scabs have fallen off. Usually, the virus is spread through droplets, and only those who come within about six and a half feet of an infectious person—what we would consider face-to-face contact—are considered at risk. In rare instances, some patients produce fine aerosols and the infection is more communicable. Examples include infectious patients with a bad cough or those with the severe, hemorrhagic form of the disease. Contaminated clothing or bedding also can spread the virus.

In past outbreaks, the average smallpox patient infected five other people. If there were an outbreak today, it is estimated that number could be as high as ten new infections per smallpox patient, since our population is considered even more highly susceptible, especially in light of increased mobility.

Symptoms of Smallpox

Smallpox has an incubation period (time from exposure to onset of illness) of seven to seventeen days. The first symptoms—high fever; fatigue, headache, severe prostration (physical collapse), and backache—usually appear twelve to fourteen days after exposure. The distinctive rash, or pox, usually breaks out two to three days later.

The rash first develops as tiny pink spots in the throat and mouth and typically spreads to the face and forearms, then to the trunk and legs. The rash is densest on the face and extremities. A distinctive characteristic of smallpox is that the rash occurs on the palms of the hands and soles of the feet. In addition, delirium strikes about 15 percent of smallpox patients. The rash grows into vesicles, or small raised bumps, that often contain fluid and within several days become filled with pus. The lesions can be extremely painful. About eight or nine days after the rash first appears, scabs start to form. Once the scabs fall off, people often are left with pitted scarring, especially on the face.

The illness caused by variola minor; which was first identified in the early 1900s, tends to be less severe, with fewer skin lesions and milder symptoms. Also, people who have some residual immunity from past vaccinations may have milder symptoms and may recover ignore quickly.

Occasionally, unusual forms of smallpox can occur: hemorrhagic smallpox and malignant smallpox. In the hemorrhagic form, bleeding into the skin and from mucous membranes occurs. In the malignant form, the rash remains flat and does not develop into the pus-filled form that is usually seen. Both forms are almost always fatal.

Smallpox as a Biological Weapon

The threat must be taken very seriously. All we need to do is look at the deadly swath smallpox has cut through history to understand how formidable a viral foe it is. And although the disease is gone, the virus that causes it still exists.

After the disease was eradicated, the WHO urged all laboratories to destroy their stocks of the *Variola virus* that causes smallpox or to transfer them to one of the two WHO-approved research labs: the Centers for Disease Control and Prevention (CDC) in Atlanta and the Institute of Virus Preparations in Moscow. Because only two labs in the world are authorized to warehouse the smallpox virus, would-be bioterrorists will likely find it exceedingly difficult to get their hands on it.

But it is suspected that nations with bioterror programs, including Iraq and North Korea, may have gained access to the virus, increasing the likelihood that it could fall into the hands of terrorists. And it is known that Russia in the 1980s secretly, and in violation of an international treaty, produced more than twenty tons of smallpox virus specifically for use in bombs and intercontinental ballistic missiles.

However, unlike other possible bioterror agents such as anthrax, botulism, and tularemia, there are no natural stores of smallpox in the soil or animals. Nor does smallpox have the ability to form spores, the hard shells that protect anthrax and botulism bacteria indefinitely in a state of suspended animation.

The technical expertise required to engineer smallpox as a bioweapon is higher than that needed to weaponize anthrax. And even if there are secret stocks stashed away, it's unlikely they would willingly be turned over to terrorists because the disease is so contagious and difficult to control. The risk is extremely high that any group or nation that unleashes this dreaded disease on the world would also inflict it on its own people.

But as unlikely as a smallpox attack seems, we must be prepared for it because the potential results could be devastating. The virus can be aerosolized, is infectious in a relatively small amount, and would quickly spread over a wide area if it is released in cool, dry weather particularly in winter. Smallpox is highly contagious, and our population is highly vulnerable to it because nobody has been vaccinated or exposed to the disease here in thirty years.

Last year, the military held an exercise called Dark Winter; in which a simulated smallpox attack was unleashed on the United States. The simulation started with 20 confirmed cases in Oklahoma City. Within just two weeks, the disease had spread like wildfire. There were 16,000 reported cases in twenty-five states, with 1,000 deaths. And smallpox outbreaks were reported in ten other countries. In another three weeks, it was estimated that there would be as many as 300,000 total victims, with 100,000 deaths. And within two months, there would be an estimated 3 million smallpox cases and up to 1 million deaths.

Although this exercise was just a simulation, the grim scenario that it predicted was a wake-up call, and steps have already been taken to fix many of the problems identified during the Dark Winter exercise. But the exercise illustrates just how dangerous a threat smallpox poses to a highly vulnerable and very mobile people.

"Biological Sucide Bombers"

It is possible. But remember, smallpox patients generally become infectious once the characteristic rash develops. By that time, they typically already have a high fever, fatigue, headache, severe prostration, and backache. In other words, they are thoroughly miserable and weak. Add to that the fact that, if they wanted the best chance to spread the disease, they would have to wait until the rash had fully erupted. It would be difficult for people with this advanced stage of the illness to walk at all, and even more difficult to walk undetected through a crowded mall or sporting event. And the infected individuals would have to get within about six feet of others to even have a chance of spreading the disease. But the unthinkable can happen.

Is Smallpox always Fatal?

No. Before the disease was eradicated, the death rate among unvaccinated persons for the most serious form, variola major, was about 30 percent. *Variola minor* kills only about 1 percent of its victims.

The vaccinia vaccine can offer protection for previously unvaccinated persons exposed to the smallpox virus, provided it's given one to three days—and perhaps even as long as four days—after exposure. People who are vaccinated soon after they are exposed may not fall ill, or if they do, the disease is usually less severe and not likely to be fatal. For this reason, early diagnosis is critical. Early vaccination is critical. Every moment counts.

At best, if you were vaccinated thirty or more years ago, you would only have partial immunity at this point. It's uncertain how much. Some experts estimate that a single dose of smallpox vaccine offers protection for probably three to five years, and not more than ten years. Since routine smallpox vaccinations ended in the United States in 1972, that means our entire population must be considered highly vulnerable.

Anyone exposed to the virus now would have to be vaccinated again, regardless of whether he had been vaccinated as a child. It is possible, however, that a revaccination for someone who was immunized as a child might boost immunity faster than a first-time vaccination. Where possible, the CDC will revaccinate health care workers and emergency personnel first so they could handle patients during the early stages of a smallpox outbreak.

Enough Vaccine

There will be by the end of 2002. The United States had 15.4 million doses of smallpox vaccine on hand at the time of the September 11 attacks. Preliminary testing suggests that the existing doses can be diluted to create five doses each, which will bring the current supply to 77 million doses. Following the terrorist attacks, the United States immediately ordered an additional 54 million doses from a Massachusetts-based company. To cover the rest of our citizens, President Bush last November signed a contract with a British firm to buy 155 million more doses, at a cost of $428 million. By the end of 2002, we will have 286 million doses on hand, enough to protect every man, woman, and child in our nation if it is ever deemed necessary.

If it weren't for the *threat of bioterrorism*, we might have been able to be forever rid of smallpox. The WHO has long advocated destroying the remaining stocks of smallpox virus, and the United States had been expected to kill off its remaining supplies this year. But the Bush administration correctly reversed that decision in the wake of the September 11 attacks. As anxious as we are to eliminate the possibility that smallpox will be reintroduced in the world, we now need the remaining stocks to help develop an arsenal of drugs to keep us safe.

Given the fact that Russian scientists succeeded in turning smallpox into a weapon of mass destruction, and the chance that rogue nations may have illegal supplies, we must act prudently. The administration wants to develop at least two antiviral drugs, a vaccine that everyone can take safely, and diagnostic tests and environmental detectors to remove this threat once and for all.

It's regrettable that, given the misery this disease has caused and the triumph of medicine over this ancient scourge, we cannot now take the final step of destroying the remaining research stocks of the virus. But clearly it would be unwise to kill off our supplies when we don't know who else might have the smallpox virus and when we still don't have a vaccine that's safe for everyone or any effective treatment once the disease strikes.

Universality of Vaccination

First of all, we don't have enough vaccine on hand right now to vaccinate everyone. But even when we do by the end of the year, the risk of side effects from vaccination outweighs the benefits of vaccination when the virus is not circulating in the population. The decision to vaccinate our population against a particular disease should be made when:

1. People have a relatively high likelihood of being exposed to the disease; in other words, the virus is circulating in the population.
2. The disease is likely to spread from person to person.
3. The effect of vaccination are not worse than the actual disease.

On the basis of our experiences from the 1950s and 1960s, we would predict that if we were to revaccinate the entire country of almost 300 million people for smallpox, at least 1,500 people would develop a serious side effect from the vaccine and at least 300 people would likely die. The most common side effects, listed from most to least serious, along with their rates of occurrence as determined by a 1968 study, include:

1. *Postvaccinial encephalitis* (12.3 cases per million vaccinations). Between eight and fifteen days after vaccination, symptoms related to inflammation of the brain develop. These symptoms include fever, headache, vomiting, drowsiness, and, sometimes, paralysis, signs of meningitis, coma, and convulsions. One in four people who developed post vaccinial encephalitis died.
2. *Progressive vaccinia* (1.5 cases per million vaccinations). This rare reaction is often fatal for those with weakened immune systems. The lesions from the vaccination fail to heal, killing adjacent skin tissue, then spreading to other parts of the skin, bones, and internal organs.

3. *Eczema vaccinatum* (38.5 cases per million vaccinations). The vaccination causes the spread of eczema, an itching inflammation of the skin with lesions.
4. *Generalized vaccinia* (241.5 cases per million vaccinations). About six to nine days after vaccination, lesions—caused by the virus being disseminated through the blood—erupt. This complication typically cleared up on its own.
5. *Inadvertent inoculation* (529.2 cases per million vaccinations). Lesions are spread from one person to another through close contact, or to other parts of the person's own body such as the face, eyelid, mouth, and genitalia. Most lesions healed on their own.

Historically, about one death occurs for every 1 million people who receive a first-time vaccination, and about one death occurs for every 4 million who are revaccinated.

These predictions may underestimate the side effects because today our population includes hundreds of thousands more people with weakened immune systems (many of whom don't know it), including people with HIV infection, people taking drugs to sup press rejection of organ transplants, and those being treated fur cancer. These people are at greater risk of serious side effects and death following vaccination. Inadvertent vaccination of some people who do not know they have weakened immune systems would likely increase the rate of serious side effects and death.

Also, the CDC recommends that an injection of a medicine known as *vaccinia immune globulin* (VIG) be given to those who develop a serious side effect. Since VIG currently is in short supply, the number of people who could die after developing a serious side effect may actually be higher than historic estimates. In addition, up to 70 percent of all children who get the smallpox vaccine have a fever of 100 degrees or higher for one or two days, and 15 to 20 percent would have fever above 102 degrees.

Routine smallpox vaccinations were discontinued thirty years ago because the risk of serious side effects from vaccination was considered to be greater than the risk of disease. Because it is far from certain that anyone will actually ever be exposed to smallpox, the risk of killing hundreds of Americans and making many more ill through vaccination outweighs the potential benefits of using the vaccine at this time. Even if smallpox were deliberately released, vaccination within three days after exposure protects almost completely against the disease. Vaccinating exposed people as late as four days after exposure may still protect against severe symptoms and death. These are the recommendations as of 2002. They may be changed as new vaccines are developed and risks of exposure to virus are modified.

Accurately Diagnosis

Because no cases of smallpox have been seen in the United States for a long time, doctors are not familiar with the illness and there is no easy test to help make the diagnosis. The initial symptoms and rash of smallpox can easily be confused with chicken pox, which is caused by a different virus and is much milder and rarely fatal.

However, there are some subtle differences that can be used to distinguish between the two diseases soon after illness strikes. The chicken pox rash itches, while the lesions of smallpox can be quite painful. The *smallpox* rash tends to appear within one or two days after the onset of illness, and the lesions all erupt and evolve at the same rate. *Chicken pox* lesions are much more superficial, pop up in groups every few days—we say they occur in "crops"—and evolve at different rates. Also, chicken pox tends not to concentrate on the face, arms, and legs, and it's almost never found on the palms of the hands or soles of the feet. If smallpox is suspected, samples of fluid from skin lesions can be tested for the virus. However; testing can be performed only in highly specialized laboratories that have adequate protective equipment and procedures in place.

Smallpox Outbreak

The CDC plan is basically quite simple: Identify those with small pox, isolate them to keep them from infecting others, and vaccinate anyone with whom they may have come into contact. The WHO has recently gone on record supporting this approach as the best method to stop a smallpox outbreak.

Once someone is diagnosed with smallpox, the key to containing an outbreak is to vaccinate everyone who has been in close contact with the victim after the onset of the rash. This strategy is called "*ring vaccination*." Close contact usually means family and friends, since generally the virus is only spread to those people who have been within about six feet of an infected person during the time that the rash is present.

Often by the time the rash has erupted and a person is infectious, he or she is severely ill and is home in bed. Once infected people are identified through their health care providers, investigators would interview them to identify others who likely were exposed to the virus in the previous three weeks. Health officials would then track down those people and immediately vaccinate them.

In the wake of the September 11 attacks, the CDC vaccinated about 140 members of special teams of disease detectives who are ready to be sent at a moment's notice to investigate a suspected outbreak anywhere in the country. In addition, the CDC has been training local and state health officials to prepare them to respond to a potential smallpox outbreak.

Vaccination Stage

If an outbreak is confirmed, the CDC can have sufficient vaccine delivered within twelve hours. Those receiving the vaccination first would likely include:

1. Anyone exposed to the initial release of the virus, provided the release was discovered during the first generation of cases at a time when vaccination would still be of benefit, or within four days after exposure.
2. Anyone who had face-to-face contact with a smallpox patient after fever set in, including all those in the same household. Although a smallpox patient is not infectious until the rash appears, this approach provides a buffer and assures that people who potentially were exposed will be vaccinated.
3. Public health, medical, and emergency personnel who care for or transport smallpox patients.
4. Lab personnel who work with specimens from patients.
5. Workers who are likely to handle infectious materials, including those involved in disposing of medical waste or linens, or disinfecting bedding and rooms.
6. Personnel involved in tracing contacts and vaccination, isolation enforcement, or law enforcement interviews of suspected smallpox patients.
7. Persons permitted to enter any facilities designated for the evaluation, treatment, or isolation of confirmed or suspected smallpox patients.
8. Persons present in a facility with a smallpox case where fine-particle aerosol transmission was likely, such as a patient with an active cough or the hemorrhagic form of the disease.

Necessity of Vaccination

If the outbreak is relatively small, it's possible that public health employees would administer the vaccinations in people's homes.

But if the outbreak grows too large, the vaccinations likely would be given in a staging area, such as a school. Mass vaccination of a community's entire population would only be undertaken if the virus were released in the air in a crowded area or if public health officials deemed that mass vaccination was needed to control the outbreak.

The CDC, however, will not force people to be vaccinated against their will. And that's the right decision. The last thing we would need is people trying to flee to avoid vaccination, under mining public confidence, creating chaos, and potentially spreading the virus even farther.

Quarantine People

Because the key to containing any outbreak is isolating those infected, special quarters might have to be set up to quarantine people if hospitals can't handle everyone who needs care. This would be done in what the CDC designates as type C isolation facilities. These would include any empty building or building not used for other purposes, such as a motel, a dedicated hospital or a separate building of a hospital, or a college dormitory. To be considered as an isolation ward, the building must have its own air-conditioning, heating, and ventilation systems that exhaust 100 percent of air to the outside through a HEPA filter, or else be at least a hundred yards from any other occupied buildings or area.

The building also must have adequate water, electricity, a dependable communications system, and the ability to support the wide range of medical care that will be needed, including IVs, oxygen machines, vital-signs monitors, ventilators to aid breathing, radiology, and basic laboratory tests. Everyone who enters a type C facility must be vaccinated.

Effective Smallpox Cure

It looks fairly promising, although more research is needed. In fact, we may already have a cure and not know it. That's because medicine has advanced tremendously since smallpox was eradicated more than twenty years ago. It seems hard to believe today, but at that time there were no drugs that could combat a virus—any virus. Now we have medicines that effectively work against AIDS, herpes, influenza, and some other viral illnesses. So it's certainly possible that antiviral medicines that have been developed for other purposes may work against smallpox. In fact, twenty-one drugs have already shown that they can kill the virus in a test tube. Whether any of those will work in people is the question we must answer. The most promising medication to date is cidofovir, or Vistide, which is used to treat eye infections caused by cytomegalovirus, a complication seen in some people with AIDS. The biggest problem is that cidofovir is only available by injection, not in pill form, which would make it difficult to use under emergency conditions. However, studies are under way to come up with at least two orally administered drugs that can treat smallpox by completely different biological means. In addition, work is ongoing to study the virus's nearly two hundred genes for clues about how strains vary. This scientific detective work could be crucial if a bioterror attack uses a genetically modified strain of smallpox to make it more dangerous.

12

BOTULISM

The toxin is widely a but the most deadly form is more difficult to obtain. It already has been used unsuccessfully in attempted bioterror attacks by a Japanese cult. Rogue nations Iraq, Iran, North Korea, and Syria are believed to have it. It would take considerable technical expertise to stabilize the toxin for aerosol release, and the toxin would lose its effectiveness rapidly because it deteriorates in bright sun. It could be used more easily to contaminate food. But cooking food at 185 degrees Fahrenheit for five minutes destroys the toxin. It would he hard to weaponize the toxin so it could be released as an aerosol. So it's not very likely to cause mass casualties in this manner. But it could be slipped into uncooked restaurant food such as condiments or a salad bar or placed in a commercial beverage. Because *botulism* causes paralysis and respiratory failure, the death rate is extremely high without the use of ventilating machines to support breathing. In cases involving contaminated food, with treatment available the death rote is about 5 percent. But it's expected that rate would be much higher if the toxin were successfully released into the air in a populated urea.

The nerve toxin that causes botulism is the most poisonous substance known to science. In theory, just a single gram of botulinum toxin released in aerosol form could kill more than a million people. That has made it one of the most researched and widely developed biological weapons on Earth. But in an odd medical twist, the toxin can cure as well as kill. It is the first biological toxin to be licensed for treatment of human disease. In the United States it is used to treat two rare eye conditions, *blepharospasm* and *strabismus*, both of which involve excessive muscle contractions. The toxin also is used to help relieve more common conditions, including *migraine headache*, *low back pain*, *stroke*, *cerebral palsy*, and *benign prostate hyperplasia*, Under the trade name Botox, it is commonly used by plastic surgeons to remove wrinkles.

What is Botulism?

Botulism is a rare, *paralyzing muscle disease* that is usually contracted from uncooked or improperly cooked food. Like anthrax, the bacterium *Clostridium botulinum* forms a hard shell, called a spore, to protect itself when the environment turns inhospitable. The spores and bacteria are normally harmless.

It is only when the bacteria grow, releasing the botulinum toxin, that they become dangerous. There are many parallels between the bacteria that cause anthrax and botulism: Both form spores and come naturally from the soil. Neither is transmissible from one person to another, and both can be spread through food and air or enter the body through wounds. And both lead to production of a toxin. Actually, there are seven distinct types of *botulinum* toxin, designated by the letters A through G. Only four of them cause illness in humans.

Botulism occurs naturally in three forms: food borne, infant, and wound. The names pretty much explain each type. On average, 110 cases of botulism are reported in the United States each year. You can't catch the disease through the skin unless there's a wound, and it doesn't spread from person to person. Although the disease is rare, it is very serious.

Symptoms of Botulism

Foodborne botulism is the form you are probably most familiar with, although it accounts for only 25 percent of cases in the United States. The toxin enters the body through contaminated food and binds to nerve endings where they join with muscles. It then blocks signals for the muscles to contract, causing *paralysis*. The first symptoms typically appear within twelve to thirty-six hours after eating contaminated food, although it can be as little as six hours and as long as ten days.

The first symptoms are double vision, blurred vision, drooping eyelids, slurred speech, difficulty swallowing, dry mouth, and muscle weakness. *Botulism* always begins in the muscles of the head, face, and neck. If untreated, the disease will then work its was down the body, paralyzing the arms, legs, trunk, and respiratory muscles.

The most common form of the disease, accounting for 72 percent of all cases, is *infant botulism*. In the United States, there are about seventy-five to one hundred cases reported each year. In these cases, normally harmless spores swallowed by an infant will germinate and release the toxin in the large intestine. This form of the disease only strikes babies under one year old and is most common at two months. Symptoms include constipation, lethargy, poor feeding, a weak cry, and poor muscle tone.

Wound botulism is extremely rare; it is usually seen in drug users. The disease acts essentially in the same way as food-born botulism. Neither infant nor wound botulism is likely to result from a bioterror attack.

Severity of Botulism

Advances in medicine over the last fifty years have cut the fatality rate for *food-borne botulism* from about 50 percent to 3 percent. The most common cause of death used to be respiratory failure. But with intensive medical and nursing care, especially the use of ventilators to help them breathe, patients can eventually recover. However, recovery takes several months because it requires the growth of new motor-nerve endings. So the demands a large number of cases would place on the health care system would be enormous. For *infant botulism*, the fatality rate is less than 2 percent, and most infants fully recover with proper treatment. Again, they may have to spend weeks or months on a ventilator.

Airborne Botlinum Toxin

Airborne release of the toxin so that it's inhaled by victims has not been studied in humans, but it's thought that the disease would progress in a similar manner to foodborne botulism. There has not been a successful aerosol release of botulinum toxin, but it's not for lack of trying. In the early 1990s, the Japanese cult Aum Shinrikyo released the toxin into the air in downtown Tokyo and at U.S. military installations in Japan during at least three failed bioterror attempts.

Botulinum toxin is colourless, odorless, and tasteless. It is easy to acquire—the Japanese cult obtained its samples from the soil locally—but difficult to turn into an effective aerosol. It is some what reassuring that, even though it was extremely well financed and had access to scientific expertise, Aum Shinrikyo was unable to turn botulinum toxin, or anthrax, into an effective bioweapon. At this point, aerosol delivery remains for the most part theoretical.

Stability of Botulinum Toxin

Using the toxin in food or beverages would be easier. The botulinum toxin is very stable on uncooked foods and untreated beverages. Contamination of a large-volume, ready-to-eat, short-shelf- life product such as a commercial beverage could produce a large number of casualties. Our water supply is considered safe from this *potential bioweapon* because the toxin can't survive standard water treatments of chlorination and aeration. Also, the amount needed to poison a reservoir would be enormous, not to mention technically difficult to produce and deliver. Cooking at temperatures above 185 degrees Fahrenheit for five minutes destroys the toxin, so food that is thoroughly cooked is considered safe. Potential targets for bioterrorists could include salad bars, condiments, arid unheated beverages such as beer wine, soda, and bottled water.

Impact of a Botulism Attack

Although the threat is considered small, it must be taken seriously because of the challenges an attack would present to our public health system. Normally about 20 percent of patients with *food borne botulism* require ventilators and extensive medical support A large out break could quickly overwhelm the number of available ventilators, critical care beds, and skilled personnel. And because recovery is very slow, the demands would continue for months.

Again, fast identification of what's happened is crucial to limiting the impact. And in our fast-moving society, that's a big challenge. Because it may be a few days before symptoms appear, a complete travel and activity history must be taken from any patient who comes down with botulism. A case in Canada clearly shows why.

Over a six-week period, twenty-eight people in two countries contracted botulism after eating an unintentionally contaminated restaurant condiment. All were initially misdiagnosed. Only after a mother and daughter were correctly diagnosed after they returned to their home more than two thousand miles from the restaurant was the botulism link discovered.

Botulism as a Bioweapon

The use of botulinum toxin as a biological weapon dates bock to World War I The Japanese biological warfare group, Unit 731, contaminated the food of Chinese prisoners in occupied Manchuria with the toxin, with predictably deadly results. The United States, fearing that Germany had weaponized botulinum toxin, prepared mare than a million doses of vaccine for Allied troops preparing to storm Normandy on D-Day. The United States and the Soviets developed the toxin as a potential weapon during the Cold War, a program that ended in the United States by executive order of President Nixon in 1969.

The Soviets, however secretly continued to test the toxin as a weapon even after signing the 1972 international treaty banning such research. Soviet scientists even attempted to splice the botulinum toxin gene into other bacteria to create a so-called *super-bug*. Four countries—Iraq, Iron, North Korea, and Syria—are believed to be developing botulinum toxin as a weapon. All are considered friendly to terrorists. We know that Iraq, leading up to the Gulf War, had loaded the toxin in a hundred bombs and thirteen Scud missile warheads.

After the war, Iraq admitted to a United Nations inspection team that it had produced a staggering nineteen thousand liters of concentrated botulinum toxin—three times more than the amount needed to kill everyone on Earth. Three thousand liters of this stock remain unaccounted for. As we look to the future, it's unsettling that Iraq viewed botulinum toxin as its main bioweapon of choice. Although the *botulinum toxin* is as lethal as any known substance, a terrorist would likely use it to create fear by inflicting few deaths that would severely destabilize a community. Even if there were no mass casualties, the use of this agent could have a major impact on our society and economy.

Vaccine

The *vaccine* that has been in use for thirty years is still considered experimental. It has been used to immunize more than three thousand lab workers in many countries. In the United States, it has been used to protect lab workers who might be exposed to the toxin and to immunize our troops during the Gulf War. The vaccine is scarce and takes several months to build up immunity. Also, it is very painful, to receive, causing considerable swelling in the arm where the injection is given. As rare as the disease is, it wouldn't make sense to vaccinate the entire population. Strange as it seems, the issue is further complicated by all of the health benefits botulinum toxin offers. Vaccinating everyone against it would deny people the advantages of medications that have been developed or will be in the future.

Effective Treatment

There is an *antitoxin horse serum* that is kept by the Centers for Disease Control and Prevention (CDC). Physicians report any botulism cases to the CDC through their state health department. Monitoring cases is essential because if someone contracts botulism from contaminated food, it's possible the poisoned food is still being served and that there could be other cases.

CDC is ready to ship the botulinum antitoxin anywhere immediately twenty-four hours a day. The serum blocks the toxin from circulating in the blood. It can stop the disease from worsening. It cannot reverse any paralysis that has already occurred. So the sooner it's administered, the better. The serum also can minimize nerve damage.

Depending on how far the disease has advanced, a patient may need to be on a ventilator to aid breathing and may require help with feeding and other support. The *paralysis*, if treated, will slowly improve, but it will take up to several months.

13

TULAREMIA

The bacterium occurs naturally in mice, squirrels, and rabbits. It was developed as a bioweapon by Japan, the United States, and the Soviet Union during World War II, and the Soviets reportedly developed antibiotic- and vaccine-resistant strains during the 1980s. The bacterium is generally unstable, and it is killed by mild heat and disinfectants. But it can survive for months in cold, moist conditions. If released in aerosol form, it could remain effective in the air for up to two hours if weather conditions are right. Tularemia bacteria ore difficult to process, and even harder to make stable enough to cause mass casualties. But *Froncisella tularensis* is one of the most infectious bacteria known, so a successfully engineered attack could infect almost every one who inhales it. With advances in treatment, the death into in the United States has plunged to less than 2 percent. Left untreated, tularemia pneumonia can kill 30 to 60 percent of the time.

Also known as *rabbit fever* or *deer fly fever*, *tularemia* is a fairly obscure and rare rural disease that made headlines in 2001 when several people on Martha's Vineyard contracted the disease while clearing brush and mowing lawns. It has, however, achieved prominence as a potential biological weapon. Why? It's one of the most infectious diseases on Earth.

WHAT IS TULAREMIA?

It's a disease caused by the bacterium *Francisella tularensis*. Inhaling as few as ten of the microscopic organisms can trigger the disease, which could lead to serious respiratory illness, including *life-threatening pneumonia*. The disease was first recognized in 1911. The number of cases in the United States has dropped dramatically, from several thou sand a year in the 1950s to fewer than two hundred annually today. Those most likely to be naturally exposed to tularemia include hunters, trappers, butchers, and farmers. Most natural cases occur in south-central and western states, especially Missouri, Arkansas, Oklahoma, South Dakota, and Montana.

The bacterium that causes the disease is not only highly infectious but also pretty resilient in the right environment. It is found naturally in water, soil, and vegetation in rural areas, and it can survive for months in cold, moist conditions. It also can remain viable for years in frozen rabbit meat.

Spreading of Tularemia

In nature, *tularemia* is primarily an animal disease carried by such small mammals as rabbits, hares, squirrels, voles, mice, and water rats. Humans most often get the disease when they are bitten by infected ticks or, less often, flies, or when bacteria enter through broken skin while they are handling infected animals or carcasses, such as skinning an infected rabbit. Such causes account for 87 percent of natural cases.

Humans also can contract *tularemia* by eating undercooked, contaminated food, such as rabbit meat; drinking contaminated water; or inhaling contaminated dust or spray. The disease is not spread from person to person.

Symptoms of Tularemia

The way the disease attacks the body depends on how it is contracted. The initial symptoms typically appear within three to five days after exposure to the bacteria but can occur from as little as one day to as long as two weeks after exposure. Symptoms include sudden onset of fever, chills, headaches, aches and pains in the muscles and joints—symptoms that are much like the common cold—a dry cough, and progressive weakness.

If the bacteria are inhaled—the method most likely to be used in a bioterror attack—the disease can rapidly progress to pneumonia, with chest pain, difficulty breathing, bloody sputum (a mixture of saliva and mucus), and, if untreated, respiratory failure and death.

In rare cases, the disease can progress to what's called "*typhoidal tularemia*," which is every bit as deadly as the version with pneumonia. With typhoidal tularemia there may be no X-ray evidence of pneumonia, and the ulcers and swollen lymph nodes that characterize the other most common forms may be absent. Instead, it's a severe blood infection that affects the entire body, and it can result in respiratory and organ failure if not caught in time.

If the disease is contracted from a bite or by handling infected materials, a red spot may appear on the skin at the point of infection and grow into an ulcer. Lymph nodes, particularly in the groin and armpits, become swollen and sore. Those who get the disease by swallowing contaminated food or drink experience inflammation in the throat and painful swelling of the lymph nodes in the neck.

Before antibiotics, up to 15 percent of those with tularemia died, and the death rate soared as high as 60 percent when pneumonia was involved. Now, with antibiotic treatment, fewer than 2 percent in the United States die after infection.

Tularemia as a Biological Weapon

Although it is important to note that it never has been used against people. It has, however, been researched as a *bioweapon*.

In the 1930s and 1940s, large outbreaks caused by contaminated drinking water erupted in Europe and the Soviet Union. As tularemia demonstrated its natural capacity to cause widespread harm, the Japanese Imperial Army's germ warfare division, Unit 731, began researching tularemia as a potential biological weapon. Some suspect that large outbreaks among Soviet soldiers during World War II were caused by deliberate release of the bacteria, but that has never been proven.

Following World War II, both the United States and the Soviet Union devoted significant resources to developing tularemia as a biological weapon. While the odds of a large-scale bioterror attack with tularemia are fairly remote, the need to be prepared is under scored by a 1970 World Health Organization (WHO) analysis.

The WHO estimated that if fifty kilograms of the bacteria that causes tularemia were released into the air over a metropolitan area of 5 million people, 250,000 would be incapacitated, and 19,000 would die. On the basis of this model, the Centers for Disease Control and Prevention has estimated that the economic toll would also be devastating: $5.4 billion for every 100,000 persons exposed.

Vaccine

Yes, but today it's only given to lab personnel who routinely work with tularemia bacteria. The *vaccine*, which is being reviewed by the U.S. Food and Drug Administration, was developed as part of the U.S. military's bioweapons research during the Cold War. It has proven effective against the disease when its contracted by handing contaminated materials in the laboratory.

However, the vaccine does not fully protect against the inhaled form of the disease. So it wouldn't make sense to vaccinate people in advance, because the inhaled form is the most dangerous and also the most likely to be used in an attack. And it takes fourteen days after vaccination for the protection it does provide to kick in, so the vaccine is not considered for use after people are exposed.

Effective Treatment

Antibiotics, started early, are very effective. *Streptomycin* is the first choice, and *gentamicin* is an acceptable alternative. Both would be given by injection, and treatment would continue for ten days. *Doxycycline* and *ciprofloxacin* (Cipro) are also effective. If *doxycycline* is used, it is continued for fourteen days to reduce the chance of relapse. *Ciprofloxacin* is given for ten days. Children would receive the same medications, but at smaller doses. *Gentamicin* is the top choice for pregnant women.

If a large-scale attack resulted in mass casualties, the first choices would be *doxycycline* and *ciprofloxacin*, taken orally, for both adults and children. The reason: It's far more efficient to dispense pills than to give large numbers of injections. Both would be avail able within hours through the National Pharmaceutical Stockpile program to supplement local needs. If there were a mass casualty, *ciprofloxacin* would be the drug of first choice for pregnant women.

Biological Weapon

It could be hard to distinguish *tularemia* from other possible causes of fever. Initially, as people start showing up with respiratory problems, it could be mistaken for influenza (the flu) or pneumonia. The tip-off would be when large numbers of previously healthy adults and children quickly become seriously ill. Even then, it would take some detective work to identify tularemia, as opposed to inhalational plague or anthrax. Plague would progress to severe pneumonia more rapidly than tularemia. And anthrax could be distinguished by the characteristic X ray showing a swelling of the lymph nodes between the lungs, in the part of the chest known as the mediastinum.

There are tests of blood, sputum, biopsy specimens, and other fluid samples to confirm tularemia. But these tests take time, and they can only be conducted at a few labs equipped to handle the highly infectious bacteria. There is no quick diagnostic test to confirm tularemia, so it will be important for public health officials to make the right call based on the symptoms they see. Early identification, rapid communication, and efficient coordination among all parts of the health care system will be crucial. Confirmation of a *tularemia* outbreak in a metropolitan area is almost a sure sign that the bacteria were deliberately released, since all previous outbreaks of the disease have occurred in rural areas.

14

Ebola and Other Viral Hemorrhagic Fevers

It would be very difficult to obtain a strain of Ebola, but it would be easier to obtain strains of arenaviruses from rodents in their natural habitat. The viruses are generally unstable, although it has been demonstrated that some can be produced in aerosol form. These viruses are highly infectious and there's a lot that science does not know about them, especially Ebola. The technology required to produce VHFs, provided the right strain could be obtained, is not considered to be highly advanced. But it would be extremely dangerous and difficult for anyone to weaponize one of the viruses. This is the biggest concern. Ebola kills up to 90 percent of those infected, depending on the strain, and there is no treatment or vaccine.

One of the things that makes the Ebola virus so frightening is how little we actually know about it. We don't know, for instance, where the disease originated or even how it is naturally spread. What we do know is that it kills with frightening efficiency—up to 90 percent of its victims die. And we know that, along with the other diseases known as viral hemorrhagic fevers, it is a bioterror threat we cannot ignore.

Viral Hemorrhagic Fevers

Hemorrhagic fever can result from infection by one of several viruses. The best-known of these viruses is the *Ebola virus*, named for the river valley in Zaire (now the Democratic Republic of Congo) where the first outbreak occurred in the mid-1970s.

The name *viral hemorrhagic fever* (VHF) actually tells you a lot about this illness. These infections cause high fever and widespread, uncontrollable bleeding. The term "*hemorrhagic*" refers to one of the diseases most gory and frightful characteristics: In its most severe form, infection causes bleeding under the skin, in internal organs, and from the mouth, nose, ears, eyes, or other body openings.

It's important to note up front that humans are not natural carriers, or hosts, for the organisms that cause VHFs. These viruses normally rely on animals or insects as disease carriers. But once an outbreak begins, the virus can be spread from person to person through direct contact with blood or body fluids.

Of the four viral families that cause VHFs, we are only going to concern ourselves with the two that are considered most likely to be used by bioterrorists. They are the filoviruses, which include *Ebola* and *Marburg hemorrhagic fevers*, and the arenaviruses, which include Lassa and South American hemorrhagic fevers.

FILOVIRUSES

When examined in a microscope, *filoviruses* appear long and stringy, or filamentous, and so they were named for their distinctive appearance.

The first virus was discovered during a fatal outbreak that occurred in 1967, when laboratory workers in Marburg, Germany, and Belgrade, Yugoslavia, developed hemorrhagic fever after handling tissues from African green monkeys that had been imported for research. The outbreak resulted in thirty-one cases and seven deaths. The specific virus was named after Marburg, and it disappeared as mysteriously as it first surfaced. Then, in 1975, a traveler who likely was exposed to the virus in Zimbabwe became ill in South Africa, passing the virus along to his traveling companion and a nurse.

In 1976, *Ebola* became the second and, to date, the only other member admitted to the filovirus family. But believe me those two are more than enough. *Ebola*, which comes in four different strains, was first identified when outbreaks of VHF occurred in northern Zaire southern Sudan. It proved even more deadly than its Marburg cousin. It killed 90 percent of the people in the Zaire outbreak, and half of those infected in Sudan.

Scientists discovered that the two outbreaks were actually caused by two different, but related, strains of the virus they came to be named *Ebola-Zaire* and *Ebola-Sudan*.

A third strain was identified in the United States in 1989, when hemorrhagic fever swept through Asiatic monkeys that had been imported for research to a Reston, Virginia, primate quarantine facility. Fortunately, this particular strain, named Ebola-Reston, did not cause illness in lab workers who were infected.

Although there were no human casualties, the outbreak occurring as it did near our nation's capital, brought Ebola to national consciousness. Its fame was further enhanced after it was chronicled in Richard Preston's best-selling book, *The Hot Zone*.

The fourth strain, *Ebola-Ivory Coast*, was identified in a patient in the Ivory Coast in 1994. This was the first time the disease had been discovered in West Africa.

Since 1976, *Ebola* outbreaks have occurred sporadically in Africa. A 1995 outbreak in Kikwit, Zaire, claimed the lives of 80 percent of the 316 people known to have contracted the disease. Unfortunately, natural Ebola virus outbreaks have continued into the twenty-first century. An extended outbreak in the Gulu district of Uganda occurred in the fall of 2000, and another claimed the lives of dozens in west-central Africa in 2001. These outbreaks are a grim reminder that we have yet to understand where this virus hides when it is not infecting people and that it remains an ongoing threat, especially in certain parts of central Africa.

Spreading of Ebola and Marburg Hemorrhagic Fevers

On the basis of what we know from outbreaks of similar diseases, we suspect that *Ebola* and *Marburg hemorrhagic fevers* are carried by animals native to Africa and the Philippines (source of the Ebola—Reston strain). And there is some evidence that certain bats native to the area where the virus has been found may be carriers, but we don't really know for sure.

Because we don't know which animals are the natural hosts for the filoviruses, we don't know with certainty how humans are infected at the start of an outbreak. We also don't know how to interrupt the chain of infection in nature, for example, as we do for West Nile virus, where we can reduce the spread to humans by spraying mosquitoes. However, we suspect that the first patient gets the disease from contact with an infected animal.

Once a person has been infected, we know how the disease spreads to others: by direct body contact with infected blood or body fluids, such as sweat, saliva or semen. That's why the out breaks we have seen in Africa have usually spread through family members and friends first. Contaminated needles or syringes also can spread the disease.

The concern among U.S. intelligence, military, and health experts is that VHF could be released in aerosol form, The Soviets, we now know, made aerosolizing Ebola a high priority in their secret bioweapons program. There even are reports that the weaponized strain they developed was named after a scientist who died of the disease.

VHF as a Bioweapon

We know that the Soviet Union, as port of its extensive Cold War bioweapons program, devoted a high priority to developing the virus as a weapon of mass destruction. Ken Alibek, a former senior scientist in the Soviet bioweapons program who defected to the United States in 1992, has stated that they even conducted experiments attempting to blend Ebola with *smallpox*.

More recently, members of the Aum Shinrikyo cult in Japan traveled to Zaire in 1992 to try to obtain Ebola virus as part of their bioterror campaign. They were unsuccessful. Although Ebola has never spread person on U.S. soil, we have experienced firsthand the horror of hemorrhagic fever. A yellow fever outbreak in 1793 in Philadelphia, then the nation's capitol, claimed the lives of 10 percent of the city's population and caused widespread panic.

Many of these viruses are naturally infectious as small particles in the laboratory. That is why they are researched only in labs that meet the most stringent biosafety standards, level 4. As we discovered in the Reston case, airborne viruses can easily infect animals, although we have not seen this happen in humans.

While some strains of virus, such as yellow fever, are quite stable in aerosol form, the filoviruses are not considered particularly stable. However, the Soviets experimented with various additives to increase the stability of the Marburg virus.

Obtaining the virus to make an aerosol would be extremely difficult, and it is thought to be beyond the capabilities of terrorist groups. And handling the virus would be extremely dangerous for anyone attempting to weaponize it.

Symptoms of Ebola

The *incubation period*, or time between exposure to the virus and the first symptoms, can be as short as several days or as long as three weeks. The first symptoms for most *Ebola* patients include high fever, headache, muscle aches, stomach pain, fatigue, and diarrhea.

Some also experience sore throat, hiccups, skin rash, red and itchy eyes, blood in vomit, and bloody diarrhea.

Within a week after the first symptoms appear, the disease can progress to chest pain, shock, and even death. The fatality rate for the disease usually ranges from 50 to 90 percent, depending on the particular strain involved. Another thing we don't know about Ebola is why some people are able to recover while others are not.

Symptoms of Marburg

After a five-to-ten-day incubation period, symptoms come on with a rush: high fever, chills, headache, and fatigue. About five days later, a red raised skin rash extending over a large area often appears, usually on the chest, back, and stomach. It's followed by nausea, vomiting, chest pain, sore throat, stomach pain, and diarrhea. The symptoms continue to grow more severe and may include jaundice, inflammation of the pancreas, severe weight loss, confusion, shock, liver failure and other internal organ problems, and the characteristic massive bleeding. About one in four people with the disease dies, although there have been outbreaks where as many as three in four die.

Treatment or Vaccine

The best that can be done is to support patients in the hospital by balancing their fluids and electrolytes, maintaining their oxy Jen levels and blood pressure, giving blood transfusions, and treating them for complicating infections.

Outbreak of VHF

In many of the outbreaks in Africa, the disease has spread quickly through hospitals because health care workers treating patients often do not wear masks, gowns, or gloves. Also, reuse of contaminated needles and syringes can spread the disease.

Isolating patients and taking necessary precautions with protective clothing and face masks, hand washing, and other barriers to spreading the virus can help contain the disease. Education is also crucial. People must be taught to prevent the disease by avoiding direct body contact with body fluids or contaminated clothing or bedding. This could reduce the spread among family members and friends.

Ebola or other VHFs as a Bioweapons

There has never been a case of the Ebola VHF in humans in the united States. But it certainly belongs on the CDC's list of highest priority bioterror agents.

1. It could potentially be highly infectious in aerosol form.
2. It causes severe illness and has a high fatality rate.
3. There is no treatment or vaccine.
4. An outbreak is likely to cause panic.
5. The Soviets made it a high priority in their clandestine offensive bioweapons program during the *Cold War*.

Given those factors, we have no choice but to consider Ebola a potential bioweapon.

What are Arenaviruses

The *arenaviruses* are a much larger family than the *filoviruses*, and their natural hosts are known: rodents. This category of viruses was first discovered in 1933, but the first member known to cause viral hemorrhagic fever, the Junin virus, was identified in 1958. Junin caused Argentine hemorrhagic fever in a large agricultural area in Argentina.

Other *arenaviruses* associated with VHF are Machupo, isolated in 1963 in Bolivia; Lassa, identified in West Africa in 1969; and the two most recent additions, Guanarito, discovered in Venezuela, and Sabia, found in Brazil. Junin, Machupo, Guanarito, and Sabia are collectively known as South American VHFs.

Spreading of Arenaviruses

Once again, we know a lot more about *arenaviruses* than we do about *filoviruses*. The *arenaviruses* are passed from generation to generation through the rodents that carry them. However, the virus doesn't appear to cause illness in the host rats and mice.

As with *Ebola virus* infections, humans are not a natural host for Lassa or the South American viruses. Humans contract the disease by direct contact with infected rodents or their urine or droppings. This can happen in various ways, including eating contaminated food, coming in contact with rodent excrement through broken skin, or whaling tiny particles contaminated by rodent urine or excrement.

Once a person is infected, the disease can spread to others in the same way as Ebola and Marburg: through direct contact with infected blood or body fluids, and through contaminated medical equipment, such as needles and syringes. Unlike Ebola, though, have been reports of airborne transmission of two of the are so wearing protective clothing, especially an N95 respirator mask, is essential when around infected patients or patients; suspected of having the infection.

Symptoms of the Lassa and South American VHFs

The incubation period between exposure to the virus and the first symptoms ranges from one to three weeks. In general, *Lassa fever* not as severe as the others. Only about 20 percent of those infected with the Lassa strain develop severe illness. The rest come down with only a mild case or show no signs of illness at all.

For those who get the severe disease, symptoms are similar to *Ebola* and *Marburg*. In addition, *Lassa fever* often involves hearing loss, tremors, and inflammation of the brain, or encephalitis, as well as serious breathing problems.

Overall, the disease kills less than 1 percent of those infected. However, among those who get the severe version, the death rate is as high as 20 percent. Pregnant women are most at risk for the severe disease.

South American hemorrhagic fevers are remarkably similar. Between one and two weeks after exposure to the virus, patients gradually experience fever, malaise, severe loss of appetite (anorexia), and fatigue. Those symptoms are soon followed by headache, back pain, dizziness, and stomach problems.

Flushing of the face and chest, tiny hemorrhages in the skin, redness of the eyes, and shock then occur. Within a week to ten days, most patients begin to improve and go on to a full recovery.

But in about 30 percent of cases, the disease progress along one of three more severe paths:

1. Pronounced bleeding from body openings, along with wide spread bruising
2. Delirium, coma, and seizures
3. A mixture of bleeding and neurological symptoms

Without treatment, 20 to 30 percent of those with South American hemorrhagic fevers die.

Treatment or Vaccine for Lassa and South American VHFs

Again, we are ahead of where we are with *Ebola* and *Marburg*. An antiviral drug called ribavirin has been successful treating Lassa fever, especially if it's used early in the illness. However, the drug is not approved in the United States for use against VHFs. And there is no vaccine.

For Junin, considered the main bioterror threat among the arenaviruses, an immune serum has been used to treat some cases.

Ribavirin also should be effective, but, again, it is not approved for this use in the United States. Likewise, a vaccine has been developed for Junin, but it also is not approved in the United States.

As with *Ebola* and *Marburg*, the staples of treatment include maintaining fluids and electrolytes and carefully monitoring blood pressure. If there were an outbreak, the same precautions would need to be taken as with *Ebola* and *Marburg*: isolating patients; wearing masks, gowns, and gloves when near someone who is infected; and making sure equipment such as needles and syringes are kept sterile.

15

Human Hepatitis

Viral hepatitis is a systemic disease primarily involving the liver. Most cases of acute viral hepatitis seen in children and adults are caused by one of the following agents: hepatitis A virus (HAV), the etiological. Agent of viral hepatitis type A (infectious hepatitis or short-incubation hepatitis); hepatitis B virus (HBV), which is associated with viral hepatitis type B (serum hepatitis or long-incubation hepatitis); and the more recently recognized non-A, non-B hepatitis viruses. The disease caused by these last viruses has been tentatively designated non-A, non-B hepatitis because no specific serological assays are currently available for their identification, and the hepatitis cannot be ascribed to either HAV, HBV, cytomegalovirus, or the Epstein-Barr virus. Non-A, non-B hepatitis accounts for most of the transfusion-associated hepatitis cases seen in the United States and a sizable portion of sporadic hepatitis. The most recently discovered hepatitis agent, the delta agent, now designated hepatitis D virus (HDV), is a defective virus which coinfects with, and requires, HBV for its own expression. Other well-characterized viruses which infrequently cause sporadic hepatitis, such as yellow fever virus, cytomegalovirus, Epstein-Barr virus (infectious mononucleosis), herpes simplex virus, rubella virus, and the enteroviruses, are discussed in other chapters of this Manual.

All of these agents produce characteristic but generally indistinguishable histopathological lesions in the liver. In individual cases, differentiation among the various types of hepatitis on clinical grounds alone is not feasible. Moreover, clinical expression of disease is extremely variable, ranging from asymptomatic (inapparent infection) to anicteric or icteric hepatitis to fulminant hepatitis and even death. The prodromal or preicteric phase is characterized frequently by fever (<39.5°C), fatigability, malaise, myalgia, blunting of olfactory and gustatory senses, anorexia, nausea, vomiting, and, especially in patients with HBV infection, maculopapular and urticarial rash, arthrologies, or polyarthritis (seen in 10 to 15% of cases). Symptoms are followed by right upper quadrant discomfort or pain associated with hepatomegaly and the appearance of dark urine and clinical jaundice. Patients with acute viral hepatitis usually recover completely. The frequency of fulminant hepatitis among icteric hepatitis B patients is probably less than 2% now that post transfusion hepatitis B has been significantly reduced by the screening of donor units for hepatitis B surface antigen. Mortality data for hepatitis A suggest that the frequency of fulminant disease in icteric cases is less than 0.5%. The importance of the delta agent is its ability to convert an asymptomatic or mild, acute or chronic HBV infection into fulminant or severe, progressive disease.

Distinguishing between viral hepatitis type A and type B assumes practical importance in light of demonstrations that hepatitis A, an infection spread by close contact via the fecal-oral route, can, on rare occasions, be transmitted parenterally, and hepatitis B, which occurs sporadically after parenteral inoculation of virus-contaminated blood or blood products, can also be spread by close intimate contact. The discovery of a unique hepatitis B antigen by Blumberg and his associates and its specific relationship to HBV infection allow serological identification of type B hepatitis cases. Now that serological markers for HA V infection are also available, all cases of acute viral hepatitis can be categorized serologically as type A or B or, by exclusion, as non-A, non-B.

Description of Agents

HAV is a non enveloped icosahedral 27- to 32-nm particle which may appear "full" or "empty" under an electron microscope. Its peak buoyant density in cesium chloride is 1.33 to 1.34 g/cm^3, but both heavier and lighter particles exist; the virus has a sedimentation coefficient ($S_{20,w}$) of 160S in a neutral sucrose solution. Studies of its nucleic acid composition indicate that HAV is an RNA virus with a linear, single-stranded genome of approximately 2.3×1.0^6 daltons. HAV has four major, structural polypeptides similar in molecular weight to those of poliovirus, and it localizes exclusively in the cytoplasm of hepatocytes. Lipid is not an integral component of HAV, which is stable to treatment with 20% ether, acid, and heat (60°C for at least

1 h); its infectivity can be preserved for years at –20°C and for at least 1 month after being dried and stored at 25°C. Viral infectivity is destroyed by autoclaving (121°C for 20 min), by boiling in water for 1 min, by dry heat (180°C for 1 h), by UV irradiation (1 min at 1.1 W), by treatment with Formalin (1:4,000 [wt/vol] for 3 days at 37°C), or by treatment with chlorine (10 to 15 ppm for 30 min) or chlorine-containing compounds (e.g., sodium hypochlorite, 10 mg/liter for 15 min). Only one serotype of HAV has been defined. It does not cross-react with HBV, and its host range appears to be limited to humans, marmosets, owl monkeys, Malaysian cynomolgous monkeys, and chimpanzees. Its properties most closely resemble those of the enterovirus subgroup of the family Picornaviridae, and it has been classified as enterovirus type 72.

Sera from patients with hepatitis B infections reveal three distinct morphological entities in varying proportions. The more numerous forms (by a factor of 10^3 to 10^4) are the small pleomorphic spherical particles measuring 17 to 25 nm in diameter (mean of 22 nm). Particle counts of 10^{13} or higher have been detected in some sera. Tubular or filamentous forms of various wealths, but with a diameter similar to that of the smaller particles, are also observed. HBV is a complex, double-shelled particle with a diameter of 42 nm. Originally designated the Dane particle, it consists of a 27-nm core surrounded by a 7- to 8-nm viral protein coat.

The nomenclature used in viral hepatitis B is shown in Table 15.1. HBV is the prototype agent for a new family of viruses designated Hepadnaviridae. The complex antigen determinant found on the surface of HBV is called hepatitis B surface antigen (HBsAg). Previous designations include Australia or Au antigen and hepatitis-associated antigen.. HBsAg on the hepatitis B virion is biochemically identical to that detected on the smaller particles and the filamentous forms and is composed of proteins, carbohydrates and lipids. Seven or more polypeptides ranging in molecular weight from 25,000 to 100,000 have been found in purified preparations of HBsAg. Two of the polypeptides, with molecular weights of 25,000 and 30..000. comprise approximately 55% by weight of the whole particle. The larger polypeptide (p30) and one or two of the larger minor components appear to be glycosylated. The 22-nm particles have an average buoyant density of 1.20 g/cm^3 in CsCl and 1.17 g/cm^3 in sucrose, a molecular weight of 3.7×10^6 to 4.6×10^6, and a sedimentation coefficient that ranges from 39 to 54S. Analysis has revealed that one antigenic specificity, designated *a*, is

Table 15.1. Nomenclature for viral hepatitis type B

Term	*Abbreviation*	*Description*
Hepatitis B virus	HBV	The 42-nm double-shelled particle that consists of a 7-nm outer shell and a 27-nm inner core. The core contains a small, circular, partially double-stranded DNA molecule and DNA polymerase activity. Originally called the Dane particle. This is the prototype for the family Hepadnaviridae.
Hepatitis B surface antigen	HBsAg	The complex antigenic determinant that is found on the surface of HBV and on the 22-nm particles and tubular forms. Formerly designated Australia (Au) antigen or hepatitis-associated antigen (HAA).
Hepatitis B core antigen	HBcAg	The antigenic specificity associated with the 27-nm core of HBV.
Hepatitis B e antigen	HBeAg	An antigenic determinant that is closely associated with the nucleocapsid of HBV.
Antibody to HBsAg, HBcAg, and HBeAg	Anti-HBs, anti-HBc, and anti-HBe	Specific antibodies produced in response to their respective antigenic determinants.

common to all HBsAg preparations. In addition, there are two sets of mutually exclusive determinants, *d* or *y* and *w* or *r*. This results in four principal subtypes of HBsAg: *adw, ayw, adr,* and *ayr*. Because of antigenic heterogeneity of the *w* determinant, there are 10 major serotypes of HBV. The predominant subtype found in North America is *adw,* followed by *ayw,* HBV subtypes do not change after infection. This fact is useful when attempting to trace an infection from one source to another. Complete (DNA-containing) HBV has a buoyant density of 1.28 g/cm³ in CsCl. The 27-nm internal core of HBV contains the hepatitis B core antigen (HBcAg), a small, circular, partially double-stranded DNA molecule, and specific DNA polymerase activity. The presence of a single-stranded region of variable length in the circular DNA molecules results in genetically heterogeneous particles with buoyant densities in CsCl that range from 1.28 to 1.38 g/cm³. The core particle contains several unique polypeptides ranging in molecular weight from 17,000 to 80,000. The hepatitis B antigen (HBeAg) has not been completely elucidated but is postulated to be an integral component of the HBV core particle, presumably existing in a cryptic form. When core particles are disrupted, a 19,000-dalton polypeptide is released which appears to contain epitopes for both HBcAg and HBeAg.

The stability of HBV does not always coincide with that of HBsAg. Immunogenicity and antigenicity are retained after exposure to ether, acid (pH 2.4 for at least 6 h), heat (98°C for 1 min; 60°C for 10 h), and up to 40 cycles of freeze-thawing. HBV is stable at 37°C for 60 min but not at 60°C for 10 h. However, inactivation may be incomplete under these conditions if the concentration of virus is excessively high. Exposure of HBsAg to 0.25% sodium hypochlorite for 3 min destroys antigenicity (and presumably infectivity). Infectivity in serum is lost after direct boiling for 2 min, autoclaving at 121°C for 20 min, or dry heat at 160°C for 1 h. Recent studies have shown that HBV is inactivated by exposure to sodium hypochlorite (500 mg of free chlorine per liter) for 10 min, 0.1 to 2% aqueous glutaraldehyde, Sporicidin (pH 7.9), 70% isopropyl alcohol, 80% ethyl alcohol at 11°C for 2 min, Wescodyne diluted 1:213, or combined β-propiolactone and UV irradiation. HBV has been shown to retain infectivity when stored at 30 to 32°C for at least 6 months, when frozen at –20°C for 15 years, and after being dried and stored at 25°C for at least 1 week.

The delta agent (HDV) is a 35- to 37-nm virus composed of a delta antigen (HDAg)-expressing core encapsidated by HBsAg and requiring the helper function of HBV to support its replication. As anticipated, HDV assumes the HBsAg subtype of the HBV infection present in the host. HDV has a small RNA genome (5.5×10^5 daltons) that is nonhomologous with HBV DNA; HDAg is a 68,000-dalton protein. The virus particle has a buoyant density of 1.25 g/cm^3 and a sedimentation coefficient intermediate between those of HBsAg and intact HBV particles. It is inactivated by Formalin under conditions similar to those which inactivate HBV. Both immunoglobulin M (IgM) and IgG antibodies to HDAg (anti-HD) can be detected during the course of HDV infection. HDV can infect only persons simultaneously or already infected with HBV.

Processing of Specimens and Environmental Control

The stability of the various serological markers for HBV and HAV eliminates the need for extraordinary collection and storage procedures if bacterial contamination is minimized. Samples can be stored at 4 to 8°C if testing is to take place within 5 to 7 days. Should longer delays be anticipated. Freezing is desirable. The addition of a bacteriostatic agent to the samples is rarcly indicated. Should this become necessary, a final concentration of 0.01 % thimerosal, 0.1% sodium azide, or gentamicin sulfate at 25 to 50 μg/ml is preferred. The use of anticoagulants or the presence of severely hemolyzed blood may occasionally cause false-positive immunoassay responses. Serum samples can be shipped at ambient temperatures if delivery is expected to be made within 48 h. Unseparated clotted blood sent in the original collection tube and received in the laboratory within 24 h does not adversely affect assay results. However, safety is best achieved by shipping samples frozen in dry ice and in doubly sealed containers, as stipulated by the Inter- state Quarantine Regulations. Although antibodies to HAV and HBV antigens are stable for years at –20°C and for at least 10 days at 37°C, repetitive freezing and thawing may lead to substantial losses in titer.

Laboratory personnel should regard all specimens collected from hepatitis patients as potentially dangerous, especially stool samples from hepatitis A candidates and blood or body fluids collected from patients with hepatitis B or non-A, non-B hepatitis. Mouth pipetting and smoking, eating, or drinking in the laboratory should be strictly forbidden. In our laboratories, needles are placed in glass bottles or metal cans and autoclaved before being discarded. All other materials are placed in discard pans and autoclaved at 121°C for 40 min. Work areas are decontaminated with 0.5% sodium hypochlorite, e.g., a 1: 10 dilution of Clorox, that is prepared fresh each month. It is used for physical cleaning of environmental surfaces, for washing hands that have been inadvertently contaminated with virus, and for decontaminating metal instruments or rotors and most labile plastics. Because hypochlorites may be corrosive to many metals, the solutions are not allowed to remain in contact with such materials for more than 5 min. Disposable gloves and gowns are worn when personnel are working with known infectious blood products or stool specimens, and hand-washing procedures are strictly enforced. Additional safety recommendations can be found in the cited literature.

Direct Examination of HAV

Immunofluorescence and immunoperoxidase staining techniques as well as thin-section electron microscopic techniques have been applied to the demonstration of HAV in hepatocytes from chimpanzees and marmosets infected with HAV. However, methods for detecting intrahepatic HA V are not applicable to clinically available specimens, for liver biopsy is rarely, if ever, performed on patients with hepatitis A. On the other hand, immune electron microscopy (IEM), which was the first method used successfully to visualize HAV, remains an important reference technique. By allowing visualization of specific immune aggregates, which are more easily detected than are monodispersed virions, this procedure enhances the sensitivity of conventional electron microscopy approximately 1,000-fold (sensitivity threshold of 10^5 to 10^6 particles per ml for IEM, compared to 10^8 to 10^9 particles per ml for electron microscopy).

For detecting HAV in stools, a 2% fecal extract is prepared from an early-acute-phase stool specimen by mixing 0.2 g of stool in 10 ml of veal infusion broth supplemented with 0.5% bovine serum albumin (BSA). The suspension is homogenized by shaking it vigorously for 5 to 10 min in a securely sealed tube containing glass beads and then is clarified by centrifugation at 4°C for 1 h at 1,000 × g. Subsequently, the supernatant fluid is passed through a series of 1.2- and 0.45-μm microfiltration membranes premoistened with 0.5% BSA to minimize virus adsorption. To reduce the hazard of infection in the laboratory, preparation of stool filtrates and of electron microscope grids must be done in an appropriate containment facility. A 0.9-ml volume of stool filtrate is incubated with 0.1 ml of a 1:10 dilution of convalescent-phase serum known to contain antibody to HAV (anti-HAV). Important controls include stool filtrates incubated with dilute alone (phosphate-buffered sa- line [PBS], pH 7.4) and with perilunes serum from a patient with hepatitis A.

After incubation for 1 h at room temperature or, alternatively, overnight at 4°C, the suspension is centrifuged at 47,000 × g for 90 min at 4°C to pellet antigen-antibody complexes. The resulting pellet is suspended in 50 μl (1 to 2 drops) of distilled water and then mixed with an equal volume of 2% phosphotungstic acid (pH 7.2). One drop of this preparation is applied to a 400-mesh Formvar carbon-coated copper grid. After 1 min, excess fluid is absorbed with filter paper, and the grid is air dried. Grids are examined by electron microscopy at a magnification of ×40,000 to 60,000, and antibody-coated aggregates or single antibody-coated particles are counted. Quantitation is achieved if a reference antiserum is used and the particles in a fixed number of grid squares are enumerated.

Because antibody-coated virus like particles other than HAV may appear in fecal extracts, a comparison between stool filtrates incubated with preillness serum and convalescent-phase serum is mandatory and preferably is done under code. This technique for visualizing HAV can be applied with minor modifications to detection of HAV in homogenates of liver (from experimental animals) and to HAV purified from liver or stool, IEM can also be modified to quantitate anti-HAV in serum.

Finally, in choosing fecal specimens for visualization of HAV, an attempt should be made to collect the earliest possible sample, for the bulk of fecal HAV shedding precedes the onset of jaundice. Because such early specimens are rarely available, detection of fecal HAV is usually not a practical clinical diagnostic technique. Instead, clinical diagnosis relies on the demonstration of a serological response.

Biological Investigations of HAV

Nonhuman primates, primarily chimpanzees and marmosets (specifically genus *Saguinus*), have been experimentally infected with fecal specimens and blood samples and have been found to develop viremia and to excrete the virus in bile and stool. Both naturally and experimentally infected animals can transmit HA V to animal caretakers. Attempts to isolate and propagate HAV in cell or organ cultures were unsuccessful until 1979, when marmoset-adapted HAV was cultivated serially in primary explant cultures of adult *Saguinus labiatus* marmoset livers and in a normal fetal rhesus monkey kidney cell line (FRhK6). A noncytopathic infection occurs with this marmoset-adapted virus.

Recently, HAV has been cultivated in vitro in a large variety of cell lines (African green monkey kidney, additional strains of fetal rhesus monkey kidney, Vero cells, the Alexander human hepatoma cell line, and human diploid lung cells). Although direct inoculation of feces and serum containing HAV has led to successful virus cultivation in vitro, isolation of HAV from clinical specimens is not sufficiently sensitive for routine diagnostic purposes. Moreover, inoculation is usually followed by a long eclipse period, rendering rapid viral diagnosis impractical. Because HAV is not cytopathic in cell culture, detection of virus in vitro requires demonstration of HAV antigen by immunoassay or immunofluorescence microscopy. A by-product of in vitro cultivation of HAV is a limitless supply of HAV antigen as a diagnostic reagent. In addition, cell culture provides a simple, sensitive method for testing the infectivity of clinical specimens. Still, cultivation of HAV remains limited to a select group of research laboratories. Practical, rapid viral diagnosis is achieved most readily by serological testing.

Serological Methods to Identify HAV and its Antibody

After the visualization of HAV in human stools and marmoset livers, a variety of in vitro assays were developed to detect HAV and its antibody, anti-HAY. Of these methods, IEM is valuable as a research and reference tool but impractical for routine diagnostic purposes. Similarly, techniques for localizing HAV in tissue have limited clinical applications. Although complement fixation is a potentially quantitative, rapid assay, its usefulness is limited by its low sensitivity, its nonspecificity, and the high frequency of anticomplementary in acute-phase hepatitis serum. The immune adherence hemagglutination assay (IAHA) was a popular approach to hepatitis A diagnostics, but, eclipsed by more sensitive, practical assays, this method is rarely used. In contrast, radioimmunoassay (RIA) and enzyme-linked immunosorbent assay (ELISA) have been applied widely and will be described in detail below.

Preparation of Reagents

The availability of commercial assays for the diagnosis of HAV infection has diminished the need for preparing reagents except in research laboratories. A combination of preparative techniques (including differential centrifugation, isopycnic ultracentrifugation, isopycnic banding in cesium chloride, rate-zonal separation in sucrose, preparative electrophoresis, gel filtration, ion-exchange chromatography, affinity chromatography, and organic solvent extraction) can be employed to purify HAV from human or nonhuman primate stool, from homogenized marmoset liver, or from cell culture for use in serological assays. If stool specimens are to be used, they should be collected as early during acute illness as possible, preferably before the onset of jaundice, after which virus excretion declines rapidly. Thus, most patients presenting with clinically obvious hepatitis A are poor sources of virus-rich stool. Success is much more likely if stool is collected from close contacts of the index case, even in the absence of clinical signs of hepatitis. Although generally impractical, this approach has been rewarding in the setting of large common-source outbreaks. Now that HAV has been grown in cell culture, an inexhaustible supply of virus antigen for diagnostic purposes has become available.

Little difficulty is encountered in obtaining a source of anti-HAV-positive serum, for the antibody response after natural infection is brisk, reaching peak titers- on the order of 1:100,000—approximately 2 to 3 months after acute illness. Furthermore, relatively high titers of serum anti-HAV are maintained years after infection. If necessary, anti-HAV can be raised in animals by immunization with purified HAV. Depending on the purity of the HAV preparation used for immunization, antisera raised in animals may have antibodies to normal liver proteins, to stool antigens, or to cell culture components. Therefore, under certain circumstances, appropriate control antigens and antisera must be used, as discussed below.

IEM

IEM can be used to detect either HAV or anti-HAV. The presence of anti-HAY in serum is determined by incubating the serum with a source of HAV containing at least 50 HAV particles per electron microscope grid square. Traditionally, a 2% stool filtrate containing HAV has been used for this purpose, although other sources may be substituted. Paired serum samples (acute phase, or preillness and convalescent phase) are tested under code to allow comparison between them and to avoid observer bias. Serum diluted 1:10 in PBS is incubated with the HAV-positive sample, and IEM is performed as described above. Particles visualized are rated on a scale of 0 to 4+, with 0 signifying no antibody and 4+ representing such heavy antibody coating that the particle surfaces are almost obscured. A rise in antibody rating of 1+, which corresponds roughly to a 10-fold increase in titer by IAHA, is considered a serological response indicative of acute infection. Preillness samples should be anti-HAV negative; acute-phase samples will be negative or barely positive (1 to 2+), and late-convalescent-phase samples, taken several weeks to months after illness, will have high antibody ratings (3 to 4+).

Difficulties in interpretation may arise when serum contains antibodies to other virus like particles that may be prevalent in stools. Generally, these particles are smaller than the 27-nm HAV particles, and an increase in antibody coating is not observed when paired sera are examined. The presence of these ubiquitous particles emphasizes the importance of testing paired serum samples simultaneously for anti-HAV and of using HAV purified from cell culture. Correspondingly, tests for HAV should include anti-HAV- negative (preinoculation or preillness) control sera.

IAHA

IAHA is a complement-dependent assay 10 to 100 times more sensitive than complement fixation and easily adapted for quantitative determinations on large numbers of samples. Although it can be used to detect HAV during virus purification, it is poorly suited to detection of HAV in unpurified liver or stool homogenates. Its widest application, therefore, has been as a serological test to detect anti-HAV. At present, IAHA is used more frequently in areas of the world other than the United States.

IAHA takes advantage of the fact that primate erythrocytes bear surface receptors for the third component of complement (C3). When a finite concentration of partially purified HAV is added to a heat-inactivated serum sample containing anti-HAV, immune complexes form. These immune complexes are incubated with a fresh complement source, followed by exposure to dithiothreitol, which stabilizes the immune complex containing C1, C4, C2, and C3 by protecting it from attack by C3b inactivator. After the addition of human group O erythrocytes, a stable complex forms between C3 and C3 receptors on the erythrocytes, resulting in hemagglutination. Hemagglutination is evaluated on a scale of 0 to 4+. All wells with hemagglutination patterns of 3+ and 4+ are considered positive if hemagglutination is absent in buffer control wells.

Because a prozone effect may be encountered, serum samples to be tested are screened at dilutions of 1:10, 1:100, and 1:1,000, and endpoint titers are determined by serial twofold dilutions within or beyond these limits. It is also important to realize that not all erythrocytes are suitable. A large number of donors must be screened; only about 20% of group O donors show marked hemagglutination activity. Once appropriate donors are found, their erythrocytes stored in Alsever solution, can be used for 2 to 4 weeks. In addition to the difficulties of maintaining a steady source of erythrocytes, IAHA has other limitations. Stool- and liver-derived HAV for IAHA must be purified by one or more biophysical techniques, and only high-titer preparations work satisfactorily. Further- more, nonionic detergents and CsCl, both used in purification of HAV, interfere with IAHA and must be removed by dialysis. Unlike other techniques for identifying anti-HAV, IAHA reactions cannot be tested for immunological specificity by blocking with purified antigen; instead, specificity must be inferred from a panel of positive and negative control sera. In addition, development of anti-HAV, as measured by IAHA, is delayed from 1 to 4 weeks compared to the earlier detection of antibody possible with other techniques. Finally, nonspecific hemagglutination is encountered when IAHA is used to measure anti-HAV in lots of immune serum globulin; treatment with kaolin has been used to minimize this problem.

RIA

Commercial RIA kits are available for diagnosing acute HAV infection, for determining the immune status of an individual to HAV after exposure, or for assessing risk in a person traveling to an endemic area or working in a high-risk environment. Advantages of RIA over other as- say systems include excellent precision, ease of performance, and unexcelled specificity and sensitivity. Limiting factors include the need for a gamma counter and the relatively short shelf life of the radiolabeled reagent.

Because anti-HAV is present in the serum of about- 40% of normal urban adults, its detection cannot, by itself, establish a diagnosis of acute disease without further Analysis. In some patients, seroconversion or a rise in anti-HAV titer may eventually result in a correct diagnosis. However, the antibody titer is already high in most patients at the time medical attention is solicited, thereby limiting the value of this approach. On the other hand, a diagnosis of acute HAV infection can readily be achieved if distinction can be made between anti-HAV of the IgM class, which appears during acute illness, and IgG antibody, which becomes pronounced during convalescence and remains elevated for years.

The HAVAB-M assay is a specific and sensitive solid-phase RIA which measures IgM-specific anti- HAV in sera. Conceptually, anti-HAV of the IgM class is removed from the serum by anti-human IgM (mu-chain specific) bound to polystyrene beads. The subsequent addition of HAV results in the attachment of the antigen to IgM anti-HAV that has combined with the solid-phase anti-human IgM. The "sandwich" is completed by adding radiolabeled anti- HAV. Specifically, serum is diluted 1:200 in PBS, and 10 μl is added to a well containing 0.2 ml of PBS (final serum concentration is 1:4,000). A goat anti-human IgM-coated bead is then added to each well, and the trays are covered with an adhesive-backed cardboard and incubated for 2 h at 24 $\pm$ 2°C. Two positive and three negative control samples are included in every test run. The beads are washed three times with distilled water, followed by the addition of 0.2 ml of HAV reagent. The covered trays are incubated for 18 to 24 h at 24 $\pm$ 2°C, after which the beads are washed again. Next, 0.2 ml of ^{125}I-labeled human anti-HAY is added to each well, the trays are covered, and incubation is continued for 4 h at 45°C. Finally, the beads are washed and counted. A cutoff value is calculated based on the positive and negative control samples and on the potency of the HAV and ^{125}I-labeled anti- HAV reagents used.

This value is determined by dividing the mean positive control counts per minute (cpm) by 10 and adding this to the mean negative control cpm. Samples with counts above the cut off are considered positive.

Other approaches have been developed to detect IgM-specific anti-HAV. These employ either concanavalin A to preferentially bind IgM (unpublished data) or staphylococcal protein A to adsorb IgM from serum samples.

The HAVAB test measures total anti-HAY antibody (IgG and IgM). It is based on the principle of competition between anti-HAV in test serum or plasma and ^{125}I-labeled anti-HAV for binding to HAV coated on a polystyrene bead. Ten microliters of each specimen is added to a microtiter well containing 0.2 ml of ^{125}I-labeled human anti-HAV. (The HAVAB protocol suggests adding the label to the 10-μl specimen, but unless the label is added promptly, the specimen may dry out or adsorb to the well.) Three negative and two positive controls are incorporated into every test run. A Formalin- and heat-inactivated HAV (primate)-coated bead is added to each well, and an adhesive cover-sealer is applied to the plate. Plates are incubated for 4 h in a 45°C water bath or for 18 to 24 h at room temperature. During this time, anti-HAV, if present in the test serum, competes with the labeled antibody for HAV binding sites on the bead. After the incubation period, the fluid contents of the wells are aspirated, and the beads are washed three times with distilled or deionized water and then transferred to counting tubes. Radioactivity bound to the beads is measured in a gamma scintillation counter, and the net (minus gamma counter background) count rate of the test sample is compared to a calculated cutoff value (one-half of the combined negative control mean and the positive control mean). Specimens with counts greater than the cutoff value are negative (no anti-HAV was available to interfere with binding of ^{125}I-labeled anti-HAV to HAV), and specimens with a count rate lower than the cutoff value are considered positive for anti- HAV. If necessary, an anti-HAV titer can be derived by testing serial serum dilutions from 1:50 to 1:3,200 (1:1,050 to 1:67,200 final concentration when the original dilution of 1:21 is taken into consideration). Dilutions are made in PBS with 0.5% BSA. The diluent is substituted for negative control serum in this situation. The anti-HAV titer is considered to be that dilution with mean cpm most nearly equivalent to, but not greater than, the cutoff value. Validity of these tests is verified by checking the N/P ratio of the negative (N) to positive (P) kit control cpm, which should be $\geq$ 5

For clinical investigators interested in preparing their own microtiter solid-phase and competitive-inhibition RIAs for detecting HAV antigen in stool filtrates or cell culture homogenates or during the various stages of purification, a simple microtiter solid-phase RIA is the most direct method available. Both methods have been used for anti-HAV.

The microtiter solid-phase RIA utilizes the sandwich principle, an arrangement created when anti- body adsorbed to an inert substance binds an antigen that is subsequently detected by the addition of specific antibody tagged with ^{125}I. To test for HAV antigen, 100 to 150 μl of a predetermined optimal concentration of high-titer anti-HAV is added to each well of a polyvinyl microtiter "U" plate. In most instances, the optimal antibody dilution lies between 1:100 and 1:10,000. The antibody is diluted in 0.05 M glycine buffer (pH 7.2) or in PBS (pH 7.4). An immune globulin preparation may also be used for coating the wells, but purified IgG is usually unnecessary. The plates are incubated overnight at 25°C (or for 2 h at 37°C). Desiccation is minimized by covering the plates with Parafilm "M" held in place by a close-fitting plastic lid. The plates are washed four times with 0.15 M NaCl containing 0.02% sodium azide and then once with PBS containing 2% fetal bovine serum. The PBS-fetal bovine serum is allowed to remain in contact with the wells at 25°C for at least 30 min to saturate any remaining binding sites. One percent BSA or gelatin works equally well. After the wash procedure, residual fluid is removed to avoid subsequent dilution of samples or reagents if the plates are to be used immediately. Conversely, the plates can be stored at 4°C for at least 6 months without significant loss of reactivity. Duplicate 25- to 75-μl samples of the material suspected of containing HAV are added to the coated wells, and incubation is continued for 2 h at 45°C by floating the Parafilm-covered plates in a water bath. Alternatively, the incubation period may be extended for 1 to 2 days at 4°C.

Appropriate controls should also be evaluated. These include HAV-positive reference material, PBS, and HAV-negative stool or gradient material, e.g., CsCl. After incubation, the wells are aspirated and washed four times with saline or PBS containing 2% fetal bovine serum or 0.05% Tween 20, and the residual fluid is removed before the addition of ^{125}I-labeled IgG containing high-titer anti-HAV. Specific activity of the antibody preparation should range from 15 to 25 μCi/μg, and a total of 200,000 to 300,000 cpm per well is employed. In general, the volume of labeled IgG added to each well should not exceed the sample volume used. To reduce background counts significantly without altering the specific binding, the label is diluted in PBS containing 50% fetal bovine serum and 10% normal human serum. Each lot of fetal bovine serum and normal human serum should be evaluated separately before use. After an additional incubation period of 1 h at 45°C (or 4 h at 25°C) and a corresponding wash step, clear adhesive tape is applied to the plate, and the wells are numbered. Each well is cut out, and residual radioactivity is measured in a gamma scintillation counter. Mean cpm of duplicate samples are compared with mean cpm for at least five negative control samples. The test for HAV is considered positive if the ratio of cpm in the sample to the mean cpm in negative control wells is ≥2.1 (sample/negative, S/N ratio). Specificity is demonstrated by blocking the binding of ^{125}I-labeled anti-HAV (IgG) with convalescent-phase, but not with preinoculation, serum from persons or experimental animals previously infected with viral hepatitis type A; by comparing results with appropriate antigen-negative controls, such as HAV- negative stool filtrates or gradient fractions; or by both methods. Similarly, hepatitis B antigens and antibodies should yield negative results.

The microtiter solid-phase RIA can also be modified in a number of ways to detect anti-HAYV Perhaps the most sensitive method is to perform an indirect blocking or inhibition assay. Briefly, 0.1 ml of a 1:10 serum dilution is preincubated for 30 min at 25°C with an equal volume of a partially purified standard preparation of HAV. Specimens containing anti-HAV combine with HAV antigen, thus reducing the amount of HAV available for binding to an antibody-coated microtiter well. Duplicate samples (75 μl) of the reaction mixture are placed in microtiter wells coated with anti-HAV, and incubation is carried out as described above. After the addition of labeled anti-HAV, the wells are counted. A reduction in residual radioactivity of ≥40% as compared with an anti-HAV-negative control sample is indicative of anti-HAV in the test serum.

A second approach substitutes serum dilutions of the sample to be tested for the precoat first antibody. If the serum contains anti-HAV, specific binding will result when HAV antigen is added to the microtiter well. Bound antigen will subsequently be detected after the addition of labeled anti-HAY to the wells. The test is considered positive if the P/N ratio is ≥2.1. A shortcoming of this procedure, however, is a reduction in sensitivity that occurs at serum dilutions of <1:200 (i.e., at higher protein concentrations).

ELISA

ELISA was first utilized in hepatitis research to detect HBsAg and anti-HBs, using the sandwich principle employed in the solid-phase RIA procedure. In this technique, an enzyme-labeled antibody is substituted for a radionuclide-labeled preparation, its presence being reflected by hydrolysis of a subsequently added enzyme substrate. The colour produced can be quantitated colorimetrically. The development of ELISA technology has spawned an interesting rivalry between the proponents of RIA and ELISA. Major objections to RIA include the expense of a gamma counter and the presence of a radiation hazard. However, in many situations the large, expensive, general-purpose gamma counters with automatic sample changers are not necessary, and hand-operated gamma counters or ^{125}I-spectrometers are similar in price to colorimeters. In addition, the radiation hazard is greatly overstated. Ten microcuries of ^{125}I (commercial kits usually contain a total of 7 to 15μCi of ^{125}I) held at a distance of 10 cm for 1 h yields an exposure rate of only 0.06 mrem (milliroentgen equivalent, man). The exposure rate is inversely proportional to the square root of the distance in centimeters. For comparison, background radiation exposure rates are about 150 mrem per year, a chest X-ray provides 10 mrem of exposure, and a gastrointestinal series provides about 8,000 mrem. Nevertheless, ELISA has a number of advantages which could recommend it over RIA, particularly in those countries where restrictive laws govern the handling of radioisotopes and the disposal of radioactive wastes. A major factor is the stability of the enzyme-antibody conjugates. These have a shelf life of months to years compared to radionuclide-labeled antibody conjugates which must be freshly prepared every 3 to 5 weeks. On the other hand, an additional manipulation is required during ELISA in that a specific substrate must be added to detect fixation of the conjugated probe. In addition, some of the enzyme substrates used are potentially carcinogenic; therefore, they should be handled with caution. Both tests are objective, quantitative, reliable, and reproducible.

Simplified, standardized commercial ELISA test kits are now available to measure anti-HAV and IgM-specific anti-HAV. These commercial tests are analogous to the commercial RIA procedures produced by the same manufacturer. In the ELISA for anti-HAV, 10 μl of each serum or plasma specimen is added with 0.2 ml of horseradish peroxidase-conjugated human anti-HAV to a microtiter well. Three negative and two positive control wells are incorporated into every test run. A HAV (human)-coated polystyrene bead is added to each well, and the microtiter plates are sealed and incubated for 3 h in a 40°C water bath or for 18 to 24 h at 24 ± 2°C. After the incubation period, the fluid contents of the wells are aspirated, and the beads are washed three times with distilled or deionized water and transferred to counting tubes. Then 0.3 ml of freshly prepared *o*-phenylenediamine substrate solution containing H_2O_2 is added to each bead-containing tube as well as to two empty tubes to serve as substrate blanks. After another 30 min of incubation, shielded from light and at 24 ± 2°C, the reaction is stopped by adding 1.0 ml of 1 N sulfuric acid to each tube. Absorbance is determined (within 2 h after the addition of sulfuric acid) in a spectrophotometer at 492 nm, and the presence of anti-HAV is determined by comparing the absorbance value of the specimen to a cutoff value (one-half the sum of the negative control mean and the positive control mean absorbance). Specimens with absorbance levels higher than the cutoff value are negative, and those with absorbance values equal to or below the cutoff are considered reactive for anti-HAV. Because of low confidence in values near the cutoff, samples with absorbances within 10% of the cutoff should be retested to confirm the initial result. Validity of the test procedure is determined by calculating the difference between the negative (N) and positive (P) control mean absorbances (N minus P), which should be 0.400. If desired, the titer of anti-HAV can be derived by testing serial dilutions of serum from 1 :50 to 1:3,200 in PBS containing 0.5% BSA. The diluent should be used as the negative control instead of the negative control se- rum. Remember that the final concentration of the sample is 21 times higher, based on the initial dilution of sample in the test well. The titer of anti-HAV is considered to be the sample dilution whose absorbance is most nearly equivalent to, but not greater than, the cutoff value.

In the ELISA for IgM-specific anti-HAV, 10 μl of a serum or plasma sample is added to 2.0 ml of normal saline; 10 μl of the diluted sample or undiluted positive and negative control samples, as described above, is then added to a microtiter well containing 0.2 ml of specimen diluent (final serum concentration of 1:4,221). Polystrene beads coated with goat antibody to human IgM (μ-chain specific) are added to each well, and the plates are incubated for 1 h at 40°C, after which the fluid contents of the well are aspirated and the beads are washed three times in distilled or deionized water. Next, 0.2 ml of a solution containing HAV (human) is added to each well, and the plates are incubated for 20 ± 2 h at 24 ± 2°C, after which the wells are aspirated and the beads are washed three more times. Then, 0.2 ml of horseradish peroxidase-conjugated human anti-HAV is added to each well. The plates are incubated at 40°C for 2 h. After the fluid is aspirated from the wells, the beads are washed and transferred to counting tubes. Freshly prepared *o*-phenylenediamine solution (0.3 ml) containing H_2O_2 is added to each tube as well as to two substrate blank tubes, and the tubes are shielded from light and incubated for 30 min at 24 ± 2°C. The reaction is terminated by adding 1.0 ml of 1 N sulfuric acid to each tube, and absorbance is read as described above. Samples with absorbance levels equal to or above the cutoff value (the sum of the negative control mean absorbance plus one-tenth of the positive control mean absorbance) are considered reactive for IgM- specific anti-HAV, while samples with absorbance values below the cutoff level are considered negative. As noted above, samples with absorbance levels within 10% of the cutoff should be retested to confirm the original result. Validity of the test requires that the difference in the mean absorbance between the positive (P) and negative (N) control samples (P minus N) is ≥ 0.400.

Commercial ELISA tests for HAV antigen are not available. To conduct such an assay, 75 μl of a predetermined optimal dilution of anti-HAV (diluted in PBS) is added to wells of polyvinyl microtiter U plates, which are incubated at 4°C for 4 h and then washed three times with PBS containing 0.05% Tween 20. The antibody-coated wells are incubated over- night at 4°C with PBS containing 1% BSA and then are aspirated and washed once before the 25-μl samples being evaluated for HAV antigen are added. The plates are covered and incubated for 15 to 18 h at 4°C; they are then washed once, and 25 μl of PBS-Tween is added and left for 15 min at room temperature. Concurrently, several wells should be incubated with preillness and convalescent-phase sera instead of PBS-Tween; these will serve to demonstrate specificity of the assay. To each well is added 25 μl of anti-HAV conjugated to horseradish peroxidase by the method of Nakane and Kawaoi. The enzyme conjugate is diluted in PBS containing 50% BSA. The plates are incubated at room temperature for 2 h and washed three times with PBS-Tween; 100 μl of freshly prepared *o*-phenylenediamine-H_2O_2 substrate is then added. After an additional 30-min incubation period in the dark, 50 μl of 2 M H_2SO_4 is added to each well to stop the reaction, and the optical density is measured at 492 nm in a spectrophotometer. Alternatively, colour can be quantitated visually on a scale of 0 to 3+. Before a sample is considered positive for HAV, specificity of the reaction should be checked by demonstrating that binding of enzyme-labeled anti-HAV is blocked with convalescent-phase but not with preillness serum from a patient with previously documented viral hepatitis type A.

Interpretation of Test Results in Hepatitis A

Because viremia is limited and because cell culture techniques are cumbersome and insensitive, demonstration of circulating HAV is difficult, impractical, and probably unwarranted for diagnostic purposes. Furthermore, because fecal shedding of virus occurs so early and because liver and bile are not routinely available as diagnostic specimens, a diagnosis of HAV infection can be made most practically by demonstrating a specific antibody response. Classically, this is done by measuring an increase in serum anti-HAV between the acute phase and convalescence; however, the interval required to demonstrate a perceptible change in antibody titer may be as long as 4 to 6 weeks. In many situations appropriately spaced serum specimens are not available, and in any event, the ability to make a diagnosis during acute illness, without having to wait for a convalescent-phase sample, is preferable. This can be achieved by demonstrating the presence of anti-HAV of the IgM class, and this approach has become the method of choice for making a diagnosis of acute viral hepatitis type A.

When testing for the presence of HAV in stool or when using HAV prepared from stool to detect anti- HAV in serum, it is important to realize that a false-positive test may result from binding of antibodies in serum to non-HAV antigens in stool. Therefore, to ensure specificity of positive results, testing with preinoculation and convalescent-phase reference reagents or appropriate blocking of the specific reaction between HAV antigen and anti-HAV conjugate with convalescent-phase, but not preinoculation, serum is necessary. Many of the difficulties encountered with false-positive results can be eliminated or minimized by using highly purified antigen, by using high-titered hyperimmune animal sera, or by using antigen and antibody reagents derived from heterologous sources. For example, nonspecific binding of antibodies in human serum to human stool components can be eliminated by using antigen derived from marmoset liver or cell culture.

Selection of the best test method in any laboratory depends on the facilities and expertise available. For detection of HAV in stool, IEM, solid-phase RIA, and ELISA are of comparable sensitivity, although the ELISA may be less specific, especially in competitive binding assays. Although RIA and ELISA are the most sensitive tests for anti-

HAV, levels of anti-HAV achieved during illness and persisting thereafter are so substantial that any of the available methods is quite satisfactory. Prevalence of anti-HAV in populations determined with all assays is virtually the same.

Direct Examination of HBV

Immunofluorescence, immunoperoxidase staining, and electron microscopy have been used extensively to examine pathological specimens and serum samples for the presence of HBV-associated antigens or particles. These procedures are not applicable to rapid, large-scale screening of HBV infections by clinical laboratories, but they have been invaluable in elucidating the biosynthetic origin of the various antigens within infected cells and for detecting the presence of immune complexes. Within the liver, HBcAg-containing particles are found predominantly in the nuclei of hepatocytes, whereas HBsAg reactivity is observed exclusively in the cytoplasm. Detection of complete virions is uncommon. The techniques employed for immunoperoxidase or immunofluorescence are well known.

As a general rule, immuno-fluorescence staining of cryostat sections is superior to the use of paraffin sections. However, cryostat sections are not always available, and cellular outlines are better resolved in uniformly cut paraffin sections than in frozen sections. Reduction of nonspecific background staining of Formalin-fixed sections can be achieved by treating tissue for 2 to 6 h at 37°C with 0.1% trypsin before staining. Muscle tissue is more resistant to digestion than liver and kidney tissue. In situations in which autologous antibody might bind to cell-associated antigen, cryostat sections can be treated for 5 min with 0.1 M glycine HCl buffer (pH 1.2) before specific antibody (labeled or unlabeled) is added.

Recently, biotin-avidin systems have been introduced for detecting viral antigens on cells or in tissues. This system offers a number of advantages over other immunodiagnostic clinical techniques: (i) the binding of avidin to biotin is extremely rapid and essentially irreversible because of the high affinity constant ($> 10^{15}$ M^{-1}) that exists between these two proteins (affinity constants of antibody for most antigens are generally 10^6-fold lower); (ii) sensitivity is enhanced, even when highly diluted primary antibody is used, because the four binding sites on avidin permit amplification of the response; (iii) Formalin-fixed, paraffin-embedded histological sections, smears, and frozen sections can be examined by conventional light microscopy; and (iv) background stain is greatly reduced. Many types of immunoperoxidase staining procedures are available. The "ABC" (avidin-biotin-labeled horseradish peroxidase complex) procedure developed by Su-Ming Hsu and his associates has been found to be highly sensitive and specific. Briefly, Formalin-fixed tissue sections are deparaffinized and hydrated through xylene and a graded alcohol series. The fixation process should employ a buffered Formalin, not exceeding 4% formaldehyde, sufficient to maintain integrity of the tissue without destroying the antigenic determinants being evaluated. Sections are incubated for 30 min with primary antibody raised against the antigen of interest. The antiserum is diluted in buffer (10 mM PBS, pH 7.6). Slides are washed for 10 min in buffer, after which diluted, biotin-labeled secondary antibody is added and incubation is continued for 30 min. This antibody, directed against the immunoglobulin class of the species used as primary antiserum, introduces biotin-labeled residues into the section at the location of the primary antibody. After washing, an avidin-biotin-labeled horseradish peroxidase complex is added which binds to the biotin-labeled secondary antibody during a 30- to 60-min incubation period. The slides are rewashed for 10 min in buffer, and tissue antigen is localized by incubating the sections for 5 min in freshly prepared peroxidase substrate solution, i.e., equal volumes of 0.1% diaminobenzidene tetrachloride prepared in 0.1 M Tris buffer (pH 7.2) and 0.02% hydrogen peroxide prepared in distilled water. The hydrogen peroxide should be prepared fresh from a concentrated stock solution. Sections are washed for 5 min in tap water, counter- stained with hematoxylin, eosin, or both, and mounted. Sections should not be allowed to dry out during the staining procedure; therefore, a humidified chamber is recommended for incubations. If antigen concentration is low, the peroxidase substrate incubation step may be lengthened to achieve maximal staining. Many mammalian tissues contain endogenous peroxidase. If this problem exists, sections may be incubated for 30 min in 0.3% hydrogen peroxide in methanol either before primary antibody is added or before the ABC reagent step. The latter point is desirable in cases where the antigenic determinant maybe destroyed by hydrogen peroxide. If nonspecific staining occurs, it can be reduced by incubating the sections for 20 min with diluted normal serum prepared from the species in which the secondary antibody is made. Excess serum is blotted from the sections before the primary antibody is added.

Cell Cultures and Animal Models of Hepatitis B

Despite strong evidence for the growth of HBV in some cell and organ cultures, serial propagation over a prolonged period of time has not been accomplished. Chimpanzees and other high-order primates are highly susceptible to experimental induction of hepatitis B, but are not routinely available. The pattern of infection is similar to that observed in humans except that the disease is milder; e.g., jaundice is rare. Nevertheless, the chimpanzee system continues to play an essential role in inactivation studies (vaccine safety, disinfection kinetics), infectivity determinations, and immunopathology.

Serological Identification of Hepatitis B Antigens and Antibodies

A number of serological techniques with varying degrees of sensitivity have been developed for the detection of HBV antigens and their specific antibodies. These methods include agarose gel diffusion, counterimmuno electrophoresis, rheophoresis, complement fixation, latex agglutination, hemagglutination, IEM, ELISA, and RIA. Each method offers certain advantages and disadvantages and differs markedly in sensitivity, specificity, simplicity, and expense from the others. A description of the ELISA, RIA, and DNA polymerase methods follows. Readers interested in detailed information concerning the purification of hepatitis B-associated antigens, production of antibodies to hepatitis B, agarose gel diffusion, discontinuous counterimmuno electrophoresis, rheophoresis, reverse passive hemagglutination, and hemagglutination. A complete list of licensed manufacturers for hepatitis B products can be obtained by writing to the Director, Bureau of Biologics, Food and Drug Administration, Bethesda, MD 20205.

Purification of Hepatitis B-associated Antigens

The widespread availability of sensitive and specific commercial kits for the detection of hepatitis B antigens and antibodies has markedly diminished the laboratory's need to prepare highly purified antigens or antibody of high specificity, affinity, and avidity. However, situations may exist, e.g., subtyping, in which reagent preparation may be desirable. For those who are interested, methods for purifying HBsAg or HBcAg are described in the 3rd edition of this Manual. The HBsAg procedure can be completed in 1 week and incorporates an acid pH step to help remove antibody specific for HBsAg and other proteins nonspecifically associated with the particles. The low pH also results in partial disruption of the 42-nm HBV particles and the tubular forms. This may have additional benefits since 27-nm particles with their higher densities (1.32 to 1.36 g/cm3 in CsCl versus 1.24 to 1.28 g/cm^3 for HBV particles) are more easily separated from the lighter 22-nm

HBsAg particles, thereby producing a final product that is less likely to be contaminated with HBcAg or to be infectious. Finally, immunogenicity and radioisotopic labeling of these purified HBsAg preparations seem to be superior to those found with other methods of purification that lack a low-pH step.

Production of Antibodies to Hepatitis B

Antisera from multiply transfused (or immunized) individuals have provided investigators with a rich source of anti-HBs of moderate titer. The major advantage of human (or chimpanzee) serum is its freedom from anti-normal human serum contaminants. Satisfactory anti-HBs for screening sera for HBsAg also has been prepared in a variety of animals, including horses, goats, guinea pigs, rabbits, mice, rhesus monkeys, baboons, and chimpanzees. However, virtually all of these preparations have detectable levels of antibody to human serum proteins, particularly albumin. These contaminating antibodies can be effectively removed by adsorption with glutaraldehyde cross-linked preparations of normal human serum.

Techniques for immunizing goats, guinea pigs, and rabbits can be found in the 3rd edition of this Manual. Antibody with excellent antigen-precipitating capacity can be prepared in goats. In general, the predominant antibody is directed against the group-specific a antigenic determinant unless specimens are obtained early in the course of immunization. Therefore, goat antiserum is better for screening than for subtyping. Antiserum prepared in guinea pigs is an excellent reagent for subtyping and for screening tests. Specificity of antisera produced against subtype *ay* is better than that produced against subtype *ad*, which suggests that the *a* subdeterminant in the *ad* preparation may be more immunogenic, resulting in a more broadly reacting antiserum. Rabbits are also useful for preparing subtyping reagents (especially against the *w* and *r* determinants) and for preparing antiserum to HBcAg.

For subtyping, monospecific IgG containing anti-*d* or anti-*y* can be obtained by affinity chromatographic techniques in which the HBV-specific antiserum is passed through a column containing the heterologous HBsAg subtype. Anti-*a* (and presumably anti-w) can subsequently be recovered from the column by elution with acid or potassium thiocyanate. More recently, commercial firms have produced a number of monoclonal antibodies of precise, quantifiable specificity and antiserum homogeneity directed against the group *a* determinant and subtypes *ad* and *ay*.

Agarose Gel Diffusion

Agarose gel diffusion is the least sensitive method available for detecting HBsAg, but it permits direct comparisons to be made between positive specimens regarding their identity, partial relatedness, or non-identity. Reinforcement of weak precipitin lines is accomplished by placing reference HBsAg-positive specimens in peripheral wells between the unknown specimens. Because HBsAg is a large molecule and diffuses slowly compared with IgG antibody, preincubation of the slides for 2 h at room temperature in a humidified chamber before the addition of anti-HBs in the center well permits precipitin lines to develop more distal to the peripheral well. The slides are observed for the development of precipitin lines at 24, 48, and 72 h.

For subtyping antigens, the same reinforcement pattern can be used. Monospecific anti-HBs containing anti-d or antibody only can be substituted for the reference antiserum to enhance specificity. Slides are observed for lines of identity or for the formation of a "spur" signifying different antigenic determinants.

Discontinuous Counterimmunoelectrophoresis

Discontinuous counterimmunoelectrophoresis or counter-electrophoresis was the most widely utilized method for detecting HBsAg before the introduction of more sensitive and specific assays such as RIA. It is still used for sub typing. The method is based on the principle that, in a relatively alkaline environment, HBsAg, which has an isoelectric pH between 4.4 and 5.2, is negatively charged and migrates in an electrophoretic field toward the anode. Conversely, IgG, being closer to its isoelectric pH, travels by electroendosmosis toward the cathode. This condition is caused by charged groups within the agarose that promote the movement of buffer through the gel toward the cathode, thereby drawing the gamma globulin with it. In this regard, agarose powders with high relative mobility (m_r) are available, making these gels especially suitable for counterimmunoelectrophoresis. A precipitin line forms when optimal concentrations of the antigen and antibody meet. By reducing the ionic strength of the agarose buffer, as compared to the buffer used in the electrophoresis chambers, a discontinuous buffer system can be prepared which enhances the movement of acidic proteins toward the anode and globulins toward the cathode. This results in increased sensitivity and speed of reaction and in sharper, more easily read precipitin lines. Plates must be examined carefully for weak precipitin lines; a magnifying lens, a darkened room, and a good oblique viewing light are helpful. Artifacts may be encountered in this system which must not be confused with a true positive reaction. These include a halo of precipitation around the well and movement of lipid over the surface of the agarose adjacent to the sample well. The latter art, fact can easily be distinguished from a true precipitin reaction within the gel by wiping the area gently with a cotton swab.

Both specificity and sensitivity depend on the use of potent precipitating antibody rendered free from anti- normal human serum by prior adsorption. False- positive results are uncommon. An imbalance between reactants, either excess HBsAg or anti-HBs, can lead to the establishment of a prozone and a false- negative result. The prozone phenomenon, which occurs in the region of HBsAg excess, can be minimized by diluting the reagent antibody in normal homologous whole serum or its globulin fraction rather than in a buffer. Two-dimensional box titrations are essential to determine the optimal concentration of reagent antibody needed in subsequent tests.

Rheophoresis

The bidimensional immunorheophoresis (rheophoresis) method relies on continuous evaporation of water through a central hole placed directly over an antigen or antibody well. Protein solutions placed in peripheral wells are transported by hydrodynamic forces to the central area of dehydration through the flow of low-ionic-strength buffer (0.01 M Tris, pH 7.6) placed external to the agarose. Sensitivity is equivalent to that of counterimmunoelectrophoresis.

Reverse Passive Latex Agglutination

The major advantages of the reverse passive latex agglutination test are speed (1 h) and simplicity, coupled with a reagent shelf life of 5 months. This makes it useful for emergency situations that do not allow ample time to evaluate a sample by standard testing procedures, provided the person performing the assay has experience in reading agglutination reactions. Unfortunately, a relatively high number of false-negative (about 5%) and false-positive (about 20%) reactions occur. The causes of these unwanted reactions include the presence of rheumatoid factor, autoimmune antibodies, heterophile antibodies, lipemic serum, albumin/globulin imbalance, electrolyte abnormalities, pH imbalance, and various drug metabolites. The false-positive rate can be cut in half by heat inactivation and by absorbing out the rheumatoid factor. All other agglutination reactions should be confirmed by a blocking test. Confirmation of weak positive reactions by another method of equivalent or greater sensitivity and specificity is essential.

Hemagglutination

Erythrocytes (human group O, turkey, sheep) are coated with purified HBsAg (passive hemagglutination) or with anti-HBs (reversed passive hemagglutination). Anti-HBs antisera are prepared in guinea pigs, sheep, chimpanzees, or horses. Detailed methodology for preparing erythrocytes and for conjugating them with IgG or HBsAg can be found in the 2nd edition of this Manual. Hemagglutination tests have been used' extensively for the detection of anti-HBs or HBsAg in human serum or recalcified plasma. Their major advantages are conservation of sera (less than 10 μl is required), the absence of a requirement for expensive equipment, rapid completion (1 to 3 h), good proficiency, and easy quantification. Disadvantages include a relatively large number of nonspecific reactions and the need for personnel experienced in hemagglutination techniques. False-positive reactions frequently occur at dilutions below 1:8, thereby reducing the sensitivity accordingly. Antibodies against ruminant IgG and Forssman antigen may be removed after heat inactivation by adsorbing the sera with uncoated erythrocytes. However, the use of control erythrocytes coated with normal immunoglobulin from the same species is preferred. False-positive reactions may also occur if the serum or buffer is contaminated with certain microorganisms or if it contains rheumatoid factor. Low dilution of high- titered antibody or HBsAg may result in a false- negative reaction because of the formation of a pro- zone. The use of a vibration-free surface is also required in the reversed passive hemagglutination test to avoid false-negative results.

ELISA

Currently, two ELISA procedures are licensed in the United States for the detection of HBsAg in human serum or plasma. These are Auszyme II (Abbott Laboratories) and Cordia H; Organon's Hepanostika has been used extensively in Europe. Specific details of the assays can be obtained from the package insert accompanying each kit. In the Auszyme II assay, HBsAg is detected by incubating 200 μl of human serum or plasma with a polystyrene bead coated with guinea pig anti-HBs. Trays are sealed with an adhesive cover, and incubation proceeds overnight (usually 20 ± 2 h) at 24 ± 2°C. Contents of the wells are aspirated, the beads are washed, and goat anti-HBs-horseradish peroxidase conjugate is added. Incubation continues at 40 ± 1°C for 1 h. During this incubation period, reactive sites on the bound HBsAg combine with the enzyme-labeled antibody to form an antibody-antigen-labeled antibody sandwich. The subsequent wash period is considered extremely critical for reducing the number of nonrepeatable false-positive reactions. The colourless enzyme substrate *o*-phenylenediamine, containing hydrogen peroxide, is added to the beads, resulting in hydrolysis and the production of a yellow-orange colour in tubes containing beads with adsorbed HBsAg and substrate. After a 30-min incubation period, the enzyme reaction is stopped by adding 1 ml of 1 N HCl or sulfuric acid. Absorbance is read at 492 nm in a spectrophotometer. Specimens giving absorbance values equal to or greater than the absorbance value of the negative control mean plus the factor 0.050 are considered positive for HBsAg. Runs are not valid if the difference between the positive and negative controls is less than 0.400 absorbance units. The Cordia H assay uses an inert disk coated with horse anti- HBs. The conjugate is horse anti-HBs labeled with alkaline phosphatase, and the enzyme substrate is *p*-nitrophenyl phosphate. NaOH is used to stop the reaction, which is read colorimetrically at 405 nm. The Hepanostika assay uses polystyrene microtiter wells coated with sheep anti-HBs. The conjugate is sheep anti-HBs IgG labeled with horseradish peroxidase. Hydrolysis of the substrate (*o*-phenylenediamine-urea peroxide) is stopped with H_2SO_4, and this timing is rather critical. The reaction is read colorimetrically at 492 nm or visually using a viewbox.

In each of these assays, appropriate positive and negative controls are added to ensure specificity of the test. The total incubation time is 20 ± 2 h for Auszyme II (although shorter times of 1.5 to 3.5 h can be employed), 2.5 h for Cordia H, and 4 h for Hepanostika. The incubation temperatures are 24, 40, and 24°C for Auszyme II (routine overnight procedure), 43°C for Cordia H, and 37 and 25°C for Hepanostika. In each assay, final readings should be made within 2 h after the addition of acid (or base for the Cordia H test). The blank should be repeated whenever prolonged interruptions occur. Care should be taken to avoid splashing specimens or reagents outside of wells or high up on the rim of the well, as such splashes may not be removed in subsequent washing and could be transferred to tubes, causing test interference. Sodium azide will poison the enzyme substrate, so care should be taken that this reagent is not present in the wash reagent.

Commercial ELISA kits are available from Abbott Laboratories for the detection of IgM-specific anti-HBc (Corzyme-M) and HBeAg and anti-HBe (Abbott HBe-EIA). The Corzyme-M test uses a modified sandwich technique to measure IgM-specific anti-HBc in serum or plasma. In this assay, an appropriately diluted specimen (10 ml in 0.5

ml of PBS) is added to a polystyrene bead on which antibody specific for human IgM (mu-chain specific) is adsorbed. This removes IgM from the patient samples. The wells are sealed with an adhesive cover, and incubation proceeds at 40 ± 1°C for 1 h. Liquid is then aspirated, and the beads are washed three times with distilled or deionized water. HBcAg is added to the bead to which IgM-specific anti-HBc may be immunologically bound, and the plates are sealed. After another incubation period at 24 ± 2°C for 20 ± 2 h, the beads are rewashed three times as previously described. In the third incubation period (40°C for 2 h), human anti-HBc, conjugated with horseradish peroxidase, is added to each well to react with any HBcAg retained on a bead by the patient's IgM-specific anti-HBc. After an incubation period of 2 h at 40°C, the liquid is aspirated from the wells. The beads are washed and transferred to tubes, and *o*-phenylenediamine solution containing hydrogen peroxide is added. After incubation at 24 ± 2°C for 30 min, 1 ml of 1 N sulfuric acid is added to each tube to stop the reaction. A yellow-orange colour develops in proportion to the amount of anti-HBc-horseradish peroxidase which bound to the bead during the previous incubation. The intensity of the colour is measured with a spectrophotometer at 492 nm. The absorbance is proportional to the quantity of IgM-specific anti-HBc present in the patient's serum. A cutoff value of 0.25 times the positive control mean plus the negative control mean is determined. Specimens giving absorbance values equal to or greater than the cutoff are considered reactive for IgM antibodies to HBcAg. If rapid results are essential, the second incubation period can be reduced to 3 h at 37°C so that the test can be completed the same day.

The tests for HBeAg and anti-HBe are performed with the same commercial kit, employ different assay principles. The HBeAg test uses a sandwich principle to measure HBeAg in serum or plasma. Plastic beads coated with human anti-HBe are added to test samples (200 ml). After the trays are sealed, the samples are incubated at 24 ± 2°C for 20 ± 2 h. The fluid is aspirated, the beads are washed, and anti-HBe (human) conjugated to horseradish peroxidase is then added. After an additional incubation period of 2 h at 40°C, the beads are washed again and transferred to tubes. *o*-Phenylenediamine solution (with hydrogen peroxide) is added, and the reactants are allowed to incubate for another 30 min at 24 ± 2°C. The reaction is stopped with 1 N sulfuric acid, and absorbancy values are determined at 492 nm. Specimens giving absorbancy values equal to or greater than the absorbancy value of the negative control mean plus a factor of 0.060 are considered reactive for HBeAg.

The ELISA is about as sensitive as RIA but is superior to reversed passive hemagglutination and reverse passive latex agglutination. Improvements in the test have reduced the number of repeatable false-positive reactions, but specificity is still lower than that observed by RIA, primarily due to technical errors. However, confirmatory testing with known negative and positive sera gives reliable final results.

RIA

The RIA technique continues to be the most sensitive and specific method available for detecting the various serological markers of hepatitis B (HBsAg, HBeAg, anti-HBs, anti-HBc, anti-HBe). The methods most commonty employed utilize the solid-phase sandwich RIA technique. The double-antibody procedure, a research tool which measures the primary interaction between antigen and antibody, can be used for studying the kinetics of this reaction. It is highly sensitive, specific, and reproducible.

The solid-phase sandwich method for HBsAg detection is comparable in sensitivity to the double-antibody RIA method, is less cumbersome, and currently is the principal system used by all commercially licensed U.S. manufacturers of hepatitis B RIA kits. These assays are capable of detecting HBsAg to a level of 0.1 to 0.5 ng/ml. The concept is identical to that described in the hepatitis A RIA section. Briefly, 200 μl of serum or plasma is added to a solid-phase system to which anti-HBs is adsorbed. The solid-phase support systems used in HBsAg assays include Sepharose, polystyrene or polyethylene beads (pearls, pellets), polystyrene tubes, or controlled-pore glass. During incubation (usually 2 h at 40 to 45°C or 18 $\pm$ 2 h at 24 $\pm$ 2°C), HBsAg forms an immunological complex with the anti-HBs at the liquid-surface interface. After the beads are washed, anti-HBs tagged with ^{125}I is added and incubation is continued (1 h at 40 to 45°C). The labeled antibody binds to any HBsAg on the support system, creating an antibody-antigen-^{125}I-antibody sandwich. The washed beads are counted in a gamma counter. By dividing cpm of the test sample (S) by the mean cpm of the negative control samples (N), an S/N ratio is calculated. Values of 2.1 and above are generally considered to be reactive (positive). Within limits, there is a direct correlation between the final count rate and the concentration of HBsAg in the specimen. However, size of the HBsAg and surface area of the support system restrict the working range to between 0.1 ng/ml and 1 μg/ml, above which level saturation (a plateau) is reached. Sera from some carriers have HBsAg concentrations above 100 μg/ml. In these situations, further dilutions are required to permit quantitation. Repeatedly positive reactions that cannot be confirmed as positive for HBsAg are highly unusual (<0.1%). Nevertheless, the seriousness of the diagnosis, with its attendant personal, social, and economic repercussions, mandates that a confirmatory test be attempted on all repeatably reactive specimens. This can be approached in one of two ways: (i) specimens may be tested with another licensed HBsAg assay, or (ii) a blocking or inhibition assay can be performed to see whether unlabeled anti-HBs will specifically inhibit the reaction. The detection of anti-HBc in the absence of anti-HBs also corroborates a positive HBsAg result.

To avoid an erroneous interpretation, the negative control samples must be comparable to the test specimen. Unfortunately, the kit "negative control values are usually 10 to 40% higher than values obtained using fresh normal (nonreactive) sera. Thus, SIN values between 1.5 and 2.1 should be viewed with suspicion whenever the kit negative control is used. Correspondingly, tests for HBsAg in cerebrospinal fluid must use "normal" cerebrospinal fluid as the control. In general, protein-deficient specimens and recalcified plasma result in higher background levels.

In an emergency, the solid-phase RIA may be modified to provide an answer within 1 h. Sensitivity is slightly less than that of the regular assay, but the number of positive specimens likely to be missed should be relatively small. To further reduce the number of false-negative values, at the expense of increasing the number of false-positive reactions, the cutoff value can be reduced to $\geq$1.5 times the negative control. Ultimately, verification of a positive reaction by the regular procedure is essential.

Solid-phase RIA for the detection of anti-HBs is similar In principle to HBsAg detection except that the specimen is incubated with polystyrene beads coated with HBsAg. Radiolabeled HBsAg is used to detect anti-HBs bound to the fixed HBsAg. Specimens with a count rate equal to or greater than 2.1 times the negative control mean cpm are reactive. Unfortunately, inter- and intralot variations in commercial anti- HBs kits and diminishing sensitivity as the kits approach their expiration date limit the usefulness of the S/N ratio for comparative purposes. With the advent of the HBsAg vaccine, the need for precision and accuracy in the assay has become more important. To permit results that can be expressed in milli-International Units milliliter (mIU/ml), Hollinger et al. have modified the RIA test. Anti-HBs concentrations are based on the First International World Health Organization (WHO) Reference Preparation for HBIG (lot 26.1.77) provided by the International Laboratory for Biological Standards, Central Laboratory of The Netherlands Red Cross Transfusion Services. An arbitrary value of 50 IU of anti-HBs has been assigned to this product; To determine mIU/ml, an S/R ratio is computed as follows:

$$\frac{\text{Sample cpm} - \text{negative control mean cpm}}{\text{Reference control cpm} - \text{negative control mean cpm}}$$

The WHO anti-HBs reference standard is diluted to contain 125 mIU/ml. Regression of the S/R ratio on the WHO reference anti-HBs concentration (0.1 to 500 mIU/ml) using the computer nonlinear regression program BMDP3R yielded the following formula:

$$\text{mIU/ml} = 130.75\ (e^{0.66765(S/R)}) - 1)$$

(reciprocal of dilution)

The lower limit of detection for anti-HBs with the current commercially available solid-phase RIA is 0.7 mIU/ml. Dilutions are usually required when anti-HBs concentrations exceed 200 mIU/ml. To conserve the WHO reference reagent, the laboratory can prepare a large batch of reference anti-HBs and determine its relationship to the WHO reference standard (diluted to 125 mIU/ml) by using the S/R ratio. The exponential value in the formula must be adjusted proportionally to agree with the new laboratory standard. For example, if the S/R relationship between the laboratory standard and the WHO reference standard is 0.500, the exponential value in the formula (0.66765) must be reduced by a factor of 0.5. Conversely, an S/R ratio of 1.5 would increase it by a factor of 1.5. Once this adjustment is made, the laboratory reference standard can be substituted for the WHO reference standard. Periodic reevaluation against the WHO standard should be performed to verify the accuracy of the laboratory standard, which is stored in small aliquots at –30°C.

The anti-HBc test is a competitive binding assay in which anti-HBc from the test sample (100 μl) competes with a constant amount of human ^{125}I-tagged anti-HBc for a limited number of binding sites found on beads coated with HBcAg. Within limits, the proportion of radioactive anti-HBc bound to the bead is inversely proportional to the concentration of anti-HBc in the test specimen. Incubation is carried out at 24 $\pm$ 2°C for 20 $\pm$ 2 h. A specimen is considered to be reactive for anti-HBc if the count rate of the unknown is less than the cutoff value (one-half the sum of the negative control mean and the positive control mean).

Other RIA tests include those for HBeAg and anti- HBe (Abbott-HBe, Abbott Laboratories). The HBeAg test uses the sandwich principle to measure HBeAg in serum or plasma (200 μl). Polystyrene beads coated with anti-HBe are added to the test samples and incubation proceeds at 24 $\pm$ 2°C for 20 $\pm$ 2 h. After the beads are washed, anti-HBe tagged with ^{125}I is added and incubation is continued at 45°C for 1 h. Count rates are increased when HBeAg is present in the serum. A positive reaction is one which is equal to or greater than the cutoff value of 2.1 times the negative control mean count rate. The test for anti-HBe is a modified competitive binding assay in which anti- HBe in the test serum (100 μl) competes with antiHBe-coated beads for a standardized amount of added HBeAg. Trays containing the beads are incubated for 20 $\pm$ 2 h at 24 $\pm$ 2°C, at the end of which time the beads are washed. If anti-HBe is present in the test sample, less HBeAg would be coupled to the bead and therefore less ^{125}I-labeled anti-HBe would bind to the bead to complete the sandwich. Within limits, the greater the amount of anti-HBe in the specimen, the lower the count taste. The count rate is compared to the cutoff value, which is one-half of the sum of the count rates of the negative control mean plus the positive control mean. Specimens with count rates equal to or less than the cutoff are considered reactive for anti-HBe, similar to the other competitive binding assays.

Tests for HBV-associated DNA Polymerase

Samples are diluted 3- to 30-fold in Tris-saline buffer (pH 7.4; 0.01 M Tris hydrochloride and 0.15 M NaCl) and are clarified at 10,000 $\times$ g for 10 min; 3.0 ml is then layered over 2.5 ml of 30% (wt/vol) sucrose containing Tris-saline buffer, 0.001 M EDTA, 0.1% 2- mercaptoethanol, and 1 mg of BSA per ml which has been precentrifuged for 10 min at 10,000 $\times$ g to remove precipitated BSA. After centrifugation for 4 h at 250,000 $\times$ g (50,000 rpm in an SW65 rotor), the supernatant fluid is removed, residual fluid is adsorbed with paper, and the pellet is resuspended in 50 μl of Tris-saline buffer containing 0.1% Nonidet P-40 and 0.1 % 2-mercaptoethanol. For evaluation of a larger number of specimens, a Beckman type 25 rotor that holds 100 1-ml tubes can be used. A 200 μl sample of a precentrifuged undiluted serum is layered over 0.5 ml of 30% (wt/vol) sucrose. The tubes are centrifuged at 24,000 rpm for 15 h at 4°C, and the pellets are recovered as described above. The removal of serum proteins from the pelleted material is essential because their presence will result in high background counts that cannot be washed away. To the resuspended pellet is added 25 ml of the reaction mixture (0.2 M Tris, pH 7.4, 0.08 M $MgCl_2$, 0.24 M NH_4Cl, 1.0 mM dATP, 1.0 mM TTP, and 0.025 nM each of [^{3}H]dCTP and [^{3}H]dGTP both at 21 Ci/mmol). The reactants are incubated at 37°C for 3 h, and then 50 ml is placed on a Whatman 3-mm paper disk, washed, and assayed for acid-precipitable H. Confirmation that the reactivity is associated with HBV is accomplished by its immunoprecipitation with anti-HBs before Nonidet P-40 detergent treatment and with anti-HBc after the addition of Nonidet P-40.

Hybridization Techniques

While hybridization procedures are beyond the scope of most clinical laboratories, an awareness of their availability is desirable.. Human serum can be analyzed for HBV DNA sequences using molecular hybridization techniques. To conduct these studies, 10 to 15 μl of serum is applied to a 0.45-μm nitrocellulose filter sheet, using 5 μl for each application. The paper should be dried between applications. Known positive and negative control samples are included in each assay. The paper is treated with 0.5 N NaOH neutralized with 1 M Tris hydrochloride (pH 7.4) containing 0.5 M NaCl and treated with proteinase K at 200 U/ml for 1 h at 37°C. After this treatment, the paper is washed twice with 0.3 M NaCl containing 0.03 M sodium citrate and then baked in vacuum at 80°C for 2 h, prehybridized, and hybridized for 24 to 36 h in a solution containing Denhardt's solution (0.1% BSA, 0.1% Ficoll, 0.1% polyvinylpyrrolidone), 0.6 M NaCl, 0.06 M sodium citrate, 0.1% sodium dodecylsulfate, 0.025 M sodium phosphate buffer (pH 6.5), 200 μg of denatured calf thymus DNA per ml, and 107 cpm of repurified recombinant cloned HBV DNA labeled with 32p (specific activity, 2×10^8 to 4×10^8 cpm/μg of DNA). After hybridization, the nitrocellulose filter is washed, dried, and autoradiographed. This method is capable of detecting visually quantities from 0.2 to 0.5 pg of HBV DNA sequences within 24 h with a 2- to 10- fold increase after 5 days of autoradiography. Similar hybridization studies can be performed using DNA extracted from small portions of frozen biopsy specimens.

Interpretation of Test Results in Hepatitis B

Successful detection of HBV serological markers depends not only on the relative sensitivity of the test procedures but also on the availability of experienced personnel who comprehend the procedure used and its idiosyncrasies and who are meticulous in their performance of the test. Provided that these conditions are met, the final evaluation and interpretation of any positive test result will be determined by the specificity of the reagents used.

It is essential for the diagnostic virologist to appreciate the difficulties encountered in preparing from human sources quality reagents that are free from contaminating human proteins. These contaminants result in the production of low concentrations of unwanted antibodies during immunization. The preparation of monospecific antibody or the prior adsorption of antisera with an insoluble immunoadsorbent prepared from normal human serum has eliminated most of these problems as has the production of HBsAg by using recombinant DNA methodology. Specificity testing with a reference antigen or antiserum also provides confirmation of the laboratory result. Additional verification of an HBsAg reaction is generated when anti-HBc is also detected.

The marked increase in sensitivity of the RIA, ELISA, or hemagglutination tests as compared to the discontinuous counter-immunoelectrophoresis test should not imply that an equivalent increase in the number of HBsAg-positive persons will be detected. Experience indicates that 70% of the positive carriers in the United States have HBsAg concentrations that are detectable by second-generation assays, leaving 30% undetectable except by the more sensitive procedures. HBsAg levels between 0.1 and 0.5 ng/ml (the lower level of sensitivity for most third-generation tests) are found in less than 5% of the HBV carriers, representing about 0.01% of the donor population. Conversely, 10 to 15% of the HBV carrier population will have HBsAg concentrations above 100 μg/ml.

The presence of HBsAg in a serum is indicative of an active hepatitis B infection, either acute or chronic. In a typical HBV infection, HBsAg will be detected 2 to 4 weeks before the transaminase level becomes abnormal and 3 to 5 weeks before the patient develops symptoms or becomes jaundiced. Anti-HBc, primarily of the IgM immunoglobulin class, usually appears when the transaminase levels begin to increase. The presence of IgM-specific anti-HBc in high titer is evidence of an acute infection. These elevated titers decline regardless of whether the disease is resolved or becomes chronic. HBsAg that persists for more than 4 to 6 months after the onset of clinical illness specifies those persons who are likely to become carriers of HBV.

The HBsAg level rises and falls steadily in acute hepatitis B, whereas the HBsAg concentration is maintained within a very narrow range in the untreated chronic carrier. Thus, it is possible to predict the potential for chronicity in patients with acute HBV disease, e.g., those who are positive for IgM-specific anti-HBc, by performing a quantitative RIA test on paired sera collected 2 to 3 weeks apart. Most comparisons can be evaluated at a dilution of 1: 10,000 (prepared in PBS containing 0.5% BSA), although some samples may require a lower dilution to obtain an SIN ratio which is on the descending slope of the RIA dose-response curve. Extreme care must be employed in preparing the dilutions. A decline in HBsAg concentration predicts eventual recovery, whereas no change in the concentration indicates that the patient has developed a persistent infection.

Because the anti-HBc test is invariably positive when HBsAg is present in a clinically ill patient, its immunodiagnostic value is to validate the HBsAg reaction. Tests with discordant results always should be repeated. However, in perhaps 5 to 10% of the acute cases of hepatitis B (especially those with fulminant disease), and more frequently during early convalescence, serum levels of HBsAg may be undetectable. Examination of these sera for IgM-specific anti-HBc may help in establishing the correct diagnosis. In the absence of anti-HBc and HBsAg, active hepatitis B disease can be excluded. In contrast, the presence of anti-HBc alone is presumptive evidence for an active HBV infection. However, this relationship is not infallible, as many patients who have recovered from hepatitis B with the development of anti-HBs and anti-HBc eventually lose one or the other.

Specimens which are HBsAg positive may be evaluated for HBeAg, anti-HBe, specific DNA polymerase reactivity, or HBV particles to determine their potential for enhanced infectivity. HBeAg-positive specimens contain high concentrations of HBV particles and are more likely to transmit hepatitis B, in contrast to anti-HBe-positive samples in which the number of HBV particles is markedly reduced.

Antibody to HBsAg usually becomes detectable 1 to 2 months after the disappearance of HBsAg. It is assumed that anti-HBs production occurs much earlier, but is not observed as a result of the formation of immune complexes with excess HBsAg. Antibody to HBsAg, with or without anti-HBc, specifies immunity against reinfection. However, RIA S/N ratios lower than 10 should be viewed with caution in the absence of anti-HBc. It has been proposed that anti-HBc, not anti-HBs, be used to determine susceptibility or immunity to HBV or to decide whether to recommend vaccination with HBsAg. Conversely, only anti-HBs develops in persons who receive the HBsAg vaccine.

Depending on the circumstances, passive transfer of anti-HBs or anti-HBc is often observed in patients receiving these antibodies during blood transfusions, after hepatitis B immune globulin administration, or in neonates of mothers with recent or past hepatitis B. Recognition of these possibilities will avoid an erroneous diagnosis of HBV infection, since passive antibodies gradually disappear over a 2- to 4-month observation period in contractus to actively produced antibodies, which are remarkably stable over many years.

Subtyping of specimens by the clinical laboratory provides additional information to the clinician or hospital epidemiologist. Since the mutually exclusive d and y subdeterminants are virus specific and not host determined. This can be helpful in determining the source of infection or can provide epidemiological evidence for the relatedness among cases.

Direct Examination of Samples for the Delta Virus

Immunofluorescence and immunoperoxidase staining of liver tissue from patients or experimental animals (chimpanzees, woodchucks with woodchuck hepatitis virus) are reliable approaches for the identification of HDAg; however, these methodologies are restricted to specialized pathology laboratories and are not practical for adoption as rapid screening tests by clinical laboratories. The methodologies for such immunohistochemical staining are routine for pathology and immunology laboratories. The anti-HD probe, to which fluorescein or peroxidase is conjugated, is usually derived from serum or plasma which contains a higher titer of anti-HD and a low level of antibody to HBcAg. Thus, the anti-HBc reactivity can be reduced substantially or eliminated entirely by dilution. However, an anti-HD-negative, anti-HBc- positive control probe should be used in parallel with the anti-HD probe to substantiate the anti-Hp specificity of staining.

Controls for other sources of nonspecificity, including autoantibodies to nuclear antigens, should be incorporated into procedures for immunohistochemical staining. With these techniques, investigators have shown that the HDAg localizes to the liver cell nucleus primarily, but rarely can be detected in the cytoplasm. Although frozen cryostat sections are preferable. HDAg is stable to Formalin fixation and paraffin embedding. Therefore. HDAg can be studied in stored paraffin-embedded tissue after digestion of the section with trypsin or pronase.

Serological Identification of Delta Antigen (HDAg) and Antibody (Anti-HD)

Antibodies to HDAg have been identified and quantitated primarily by RIA or ELISA. The major limitation to wider availability of these diagnostic techniques has been a shortage of HDAg-containing clinical material for use in immunoassays. As discussed below, however, the ability to perform these tests is increasing among diagnostic laboratories.

Purification of HDAg and Preparation of Antibodies

Tests for anti-HD require a source of HDAg. Previously the HDAg employed for these purposes had been obtained from HDAg-positive human (postmortem) or chimpanzee liver or serum. More recently, a more reliable source, HDAg-positive woodchuck liver, has become available. HDAg can be extracted from human live with strong dissociating agents such as 6 M guanidine hydrochloride or 8 M urea. When liver tissue is obtained from experimentally infected chimpanzees or woodchucks at the peak of intrahepatic HDAg expression, i.e., before the appearance of anti- HD, HDAg can be harvested by simple aqueous extraction. Although HDAg is rarely detectable in patients with acute delta infection, occasionally sufficient HDAg is present in serum to serve as a source of antigen for diagnostic testing.

Serum containing anti-HD can be obtained from patients or experimental animals with acute or chronic delta infection. As mentioned above, anti-HBc reactivity, invariably present in anti-HD-positive serum, is often low in serum samples with a high level of anti-HD activity. Residual anti-HBc activity can be diluted out, and HBsAg is removed by ultracentrifugation.

RIA

Because the availability of HDAg remains limited, the only practical approach to routine laboratory immunodiagnosis is commercial immunoassays in which HDAg is derived from the livers of woodchucks with experimental woodchuck hepatitis virus and delta infection. A commercial RIA for anti-HD has recently become available for research purposes. The test is a competitive-binding RIA in which anti-HD in the test serum competes with ^{125}I-labeled human anti-HD for woodchuck liver-derived HDAg coating a solid-phase polystyrene bead. In this assay, 100 μl of ^{125}I-labeled anti-HD and 100 μl of serum or plasma to be tested are delivered to a microtiter well and an HDAg-coated bead is added. Three negative control and two positive control wells are incorporated into each test run. Microtiter plates are incubated for 20 ± 2 h at 24 ± 2°C. After incubation, liquid is aspirated from the wells. The beads are washed three times with distilled or deionized water and transferred to counting tubes, and the tubes are assayed in a gamma scintillation counter. The net cpm (minus background) for the test sample is compared with a calculated cutoff (0.4 of the mean negative control counts plus 0.6 of the mean positive control counts). Samples with a count rate equal to or below the cutoff are considered reactive for anti-HD; counts above the cutoff range are considered negative. Validity of the test requires that, for each run, the ratio of the mean negative (N) control counts to the mean positive (P) control counts should be 2: 4.0. The titer of anti-HD can be determined by testing serial dilutions of serum or plasma and sélecting the dilution which yields counts closest to, but not greater than, the cutoff value as the anti-HD titer.

Delta antigen can be detected by solid-phase sandwich RIA, as described by Rizzetto et al. The test is based on binding of HDAg in serum to anti-HD adherent to a solid phase, followed by incubation with an ^{125}I-labeled purified IgG anti-HD probe. Because HDAg always circulates within an HBsAg-encapsidated virion, detection of HDAg requires detergent (0.5% Tween-80, Nonidet P-40, or deoxycholate) disruption of the virion to expose the internal, otherwise sequestered antigen. Adding to the difficulty of HDAg detection is the transience of delta antigenemia, occurring during early infection. This simple solid-phase RIA for HDAg can be modified for detection of anti-HD in an RIA blocking assay. A standardized quantity of HDAg is added to beads to which anti-HD is adsorbed. Nonreactive anti-HD from serum or plasma competes with a constant amount of ^{125}I-labeled anti-HD for the HDAg immunologically bound to the beads. A reduction in the count rate of $\geq$50% is evidence for the presence of anti-AD in the sample. Measurement of anti-AD titers is achieved by diluting serum and determining that dilution which inhibits 50% of binding as compared to negative control sera.

ELISA

A commercial enzyme immunoassay for anti-HD, based on the configuration described above for the commercial RIA, is being developed and will be available shortly.

New Diagnostic Approaches

IgM anti-HD. Acute delta hepatitis infection (simultaneous with acute HBV infection or acute delta hepatitis superinfection of an HBV carrier) is accompanied by an early anti-HD response predominantly of the IgM class. An antibody class-capture solid-phase RIA for IgM anti-HD has been described in which antibody to human IgM (mu-chain specific) is bound to a solid phase and test serum is added. If the test serum contains IgM anti-HD, it will bind to the solid phase. When HDAg is added subsequently, it binds to the IgM anti-HD. The sandwich is completed when the labeled probe, ^{125}I-conjugated IgG anti-HD, binds to the HDAg. The methodology for this technique is analogous to that for the detection of IgM-specific anti-HBc and IgM-specific anti-HAV.

cDNA probe. Because HDAg is rarely detectable in serum, even when the serum is known to be infectious, and because anti-HD is present indefinitely in patients with chronic delta hepatitis infection, liver biopsy with immunohistochemical staining for intra-hepatic HDAg is the only reliable way to demonstrate ongoing delta hepatitis replication. A noninvasive approach for detecting HDV RNA in serum has been made possible with the recent availability of cloned cDNA probes. The cDNA probe for HDV-associated RNA is incubated with serum, and HDV RNA is identified by dot-blot hybridization analysis. This technique remains limited to a small number of research laboratories but holds promise as a simple, noninvasive test for the diagnosis of chronic delta hepatitis infection.

Interpretation of Test Results in Delta Hepatitis

Infection with HDV can occur in the presence of acute or chronic HBV infection. The duration of HBV infection determines the duration of delta infection. When acute delta and HBV infection occur simultaneously, clinical and biochemical features may be indistinguishable from those of HBV infection alone. The presence of delta hepatitis infection can be identified by demonstrating intrahepatic HDAg or, more practically, an anti-HD seroconversion (a rise in titer of anti-HD or de novo appearance of IgM-specific anti-HD, which is briefly detectable, if at all). Because IgM-specific anti-HD is transient and IgG anti-HD is often undetectable once HBsAg disappears, retrospective serodiagnosis of acute, self-limited, simultaneous HBV and HDV infection is difficult.

In contrast to patients with acute HBV infection, patients with chronic HBV infection can support HDV replication indefinitely. This can happen when acute HDV infection occurs in the presence of a nonresolving acute HBV infection. More commonly, acute HDV infection becomes chronic when it is superimposed on an underlying chronic HBV infection. In such cases, the delta superinfection appears as a clinical exacerbation or an episode resembling acute viral hepatitis in someone already chronically infected with HBV. In the past, events resembling acute hepatitis in an HBV carrier or a patient with chronic hepatitis B were attributed to superimposed non-A, non-B hepatitis or to the natural history of the disease. A proportion of such episodes, however, represents acute superinfection with HDV.

When a patient presents with acute hepatitis and has HBsAg and anti-HD in his serum, determination of the class of anti-HBc is helpful in establishing the relationship between infection with HBV and HDV. Although IgM-specific anti-HBc does not distinguish absolutely between acute and chronic HBV infection, its presence is a reliable indicator of recent infection and its absence is a reliable indicator of infection in the remote past. In simultaneous acute HBV and HDV infections, IgM-specific anti-HBc will be detectable, whereas in acute delta hepatitis infection superimposed upon chronic HBV infection, anti-HBc will primarily be of the IgG class.

As noted above, cDNA tests for the presence of HDV-associated RNA will be useful in: the future for determining the presence of ongoing HDV replication and relative infectivity. Currently, probes for this marker are restricted to a limited number of research laboratories.

Non-A, Non-B Hepatitis

Non-A, non-B hepatitis accounts for approximately 90% of the cases of transfusion-associated hepatitis seen in the United States and may be responsible for up to 20% of endemic and sporadic hepatitis. Transmission mechanisms are similar to those for hepatitis B. The incubation period ranges from 5 to 10 weeks, although longer and shorter periods have been observed. Clinical characteristics are similar to those observed after hepatitis B except that chronic liver disease may be more prominent. Experimental trans- mission of non-A, non-B hepatitis agents has been accomplished in chimpanzees. Patients with or without biochemical evidence of liver disease can transmit this disease, although the risk is significantly greater when the donor has alanine aminotransferase abnormalities. Viremia appears to precede the liver disease by at least 12 days. Evidence for at least two distinct non-A, non-B hepatitis agents has been obtained.

Diagnosis of this entity is by exclusion of HAV, HBV, cytomegalovirus, and Epstein-Barr virus in a patient who presents biochemical evidence of hepatitis. Several investigators have reported the detection of new antigens or antibodies, and even HBV-like serological markers, in the sera and liver biopsies of these patients, but none of these putative virus markers has yet been proven to be specific for non-A, non-B hepatitis. Recently, Seto et al. reported the detection of particle-associated reverse transcriptase activity in two plasma-derived products and four human sera which previously had been shown to transmit non-A, non-B hepatitis to humans and chimpanzees. In addition, sera collected from 12 patients with acute or chronic non-A, non-B hepatitis also were positive, whereas reverse transcriptase activity was found in only 2 of 48 controls.

Unexpectedly, neutralization of infectivity with presumably convalescent- phase serum or immune globulin has been unsuccessful. In another recent development, Prince et al. reported the successful isolation of infectious membrane-coated virus particles from chimpanzee liver cell cultures inoculated with sera previously implicated in transmission of non-A, non-B hepatitis. The particles were 85 to 90 nm in diameter with a core of 40 to 45 nm. Cell homogenates from infected cultures were capable of causing hepatitis in chimpanzees. Chloroform extraction of this material appeared to inactivate the virus. If each of the above findings can be confirmed, progress toward the eventual control of non-A, non-B hepatitis will have been advanced.

INDEX

C